# DEPENDABLE ENGINES
## THE STORY OF PRATT & WHITNEY

# DEPENDABLE ENGINES
## THE STORY OF PRATT & WHITNEY

## MARK P. SULLIVAN
### FOREWORD BY STEVEN F. UDVAR-HAZY

**Ned Allen**
*Editor-in-Chief, LIBRARY OF FLIGHT*

LIBRARY OF FLIGHT

Much of the material for this book was taken from company publications such as *Beehive, The Powerplant,* and *Pratt & Whitney News* and *Directions*. In addition, many press releases, fact sheets, and other publications proved valuable even if they are not specifically cited.

American Institute of Aeronautics and Astronautics, Inc., Reston, Virginia

1 2 3 4 5

**Library of Congress Cataloging-in-Publication Data**
Sullivan, Mark P., 1945-
  Dependable engines : the story of Pratt & Whitney / Mark P. Sullivan ;
foreword by Steven F. Udvar-Hazy.
      p. cm. -- (Library of flight)
  Includes bibliographical references and index.
   ISBN 978-1-56347-957-1 (hardcover : alk. paper) -- ISBN 978-1-56347-958-8
(pbk. : alk. paper)
  1.  Pratt & Whitney Aircraft Group--History. 2.
Airplanes--Motors--History. 3.  Jet engines--History.  I. Title.
   TL724.5.P7S85 2008
   629.134'3509--dc22
                              2008025588

Book design by Pre-PressPMG, 4 Collins Avenue, Plymouth, MA 02360

# FOREWORD

The story of Pratt & Whitney is a kaleidoscope of aviation passion, dreams, opportunities, vision, and courage, fused with the impact of turns and twists of history. Frederick Rentschler's single-handed commitment to visualizing, designing, and building dependable engines created Pratt & Whitney's unique capability to advance America's leadership in aircraft engines. This legacy has embraced many decades, beginning in 1925. The Pratt & Whitney R-1340 Wasp, the Hornet, the R985, the Twin Wasp R-1830, the R-2000, and the R-2800 Double Wasp reciprocating engines progressively raised the performance bar for both commercial and military aircraft applications in the period leading up to World War II.

On a personal level, my first flying experience was in my native Hungary during the summer of 1955 on a Russian-made LI-2 (built under license from the Douglas Aircraft Company) and powered by a Russian copycat of the R-1830. What an unforgettable thrill for a nine-year-old boy! Ironically, Pratt & Whitney enabled my escape to freedom to the West on April 10, 1958, onboard an R-2800–powered Douglas DC-6 airliner, and then my first journey to the United States occurred from Stockholm, Sweden, via Copenhagen (Kastrup)–Prestwick (Scotland) and on to New York on July 4, 1958, onboard an SAS DC-6B powered by R-2800-CB16 engines that glowed through the Atlantic night sky. The sound of four R-2800s at full takeoff power undeniably captivated the emotion of all aviation enthusiasts; I was no exception.

As a 14-year-old teenage airplane addict, I spent many afternoons after school at New York's La Guardia Airport, wandering around the terminal viewing terraces and hangars. The most memorable impressions were undoubtedly the DC-3s, DC-4s, Martin 404s, Convair 340/440s, DC-6s, and DC-6Bs, all powered by Pratt & Whitney engines and operated by airlines that included Northeast, TWA, Eastern, American, United, Capital, Allegheny and Mohawk—what memories indeed that fed the inner instincts to pursue aviation as a lifelong adventure.

Fast forward to September 20, 1973, 35 years ago, International Lease Finance Corporation (ILFC) begins a new era in commercial aviation by its first operating lease of a Douglas DC-8-51, leased to Aeromexico, powered by Pratt & Whitney JT3D-3B turbofan engines. This was our very first owned jet, and how proud we were that September day in 1973. Many Pratt & Whitney–powered jets followed in the 1970s and paralleled ILFC's growth and success, including dozens of Boeing 720-B, 707, and DC-8-50 and 60 series. Then on April 15, 1977, we received our first brand new ILFC jet, a Boeing 737-204 Advanced, to be leased to Britannia Airways, powered by JT8D-15A engines. Then came more B737-200s, a year later our first new

B727-214 powered by PW JT8D-9As, then new MD-82s two years later powered by JT8D-217As, and soon after our first B747-200B, powered by JT9D-7Qs.

The 1980s saw our first B757s with PW 2037s and the 1990s with PW 4000s powering our new MD-11s, B767-300ERs, and our first A330-300s in 1994. At each important milestone, Pratt & Whitney's logo with its Eagle and "Dependable Engines" was the fitting and proud insignia on the engines that powered these magnificent flying machines.

The mid-l990s also saw another key development in the unique relationship between Pratt & Whitney and ILFC. The International Aero Engines (IAE) V2500 engine became a trusted powerplant on our growing fleet of Airbus A320s, A319s, and A321s. In 2008, ILFC is by far IAE's largest customer with more than 500 engines delivered or in the order backlog.

As we look to the future, Pratt & Whitney is designing, testing, and developing a new family of GTF jet engines, which once again will reshape the fabric and future of the commercial aviation business in the coming decade, just as the Pratt & Whitney R-1830 did in the 1930s, the Pratt & Whitney R-2800 did in the 1940s, the JT3 did in the late 1950s, the JT8D did in the 1960s, the JT9D did in the 1970s, and the PW2000 and PW4000 did in the 1980s and 1990s.

Pratt & Whitney's unique history and contributions to civilian and military aviation could fill hundreds of volumes but are instead so elegantly condensed and portrayed in this book. The impact of Pratt & Whitney has touched the lives of all who are engaged in the world of aviation and flying. This book so vividly captures the colorful history of the men, women, and machines that led us on this path of technological advancement for nearly 90 years. We respectfully salute their accomplishments.

As we approach the next decade, the blue skies over Connecticut will once again reflect the innovation, courage, and talents of the people who make Pratt & Whitney a very special contributor to the impact of aviation on our planet. This introduction is a tribute to those at Pratt & Whitney who have made this history so rich and forever exciting.

Steven F. Udvar-Hazy
Chairman of the Board
Chief Executive Officer
and Founder of International Lease Finance
Corporation

# CONTENTS

preface                                                                                          viii

acknowledgments                                                                                    ix


chapter 1        Mr. Rentschler Comes to Hartford                                                   1

chapter 2        Clouds on the Horizon                                                             15

chapter 3        50,000 Airplanes                                                                  25

chapter 4        What Do We Do Now?                                                                33

chapter 5        Forging Ahead ... Right into Space                                                47

                 Timeline                                                                          62

chapter 6        Engines That Just Kept Going and Going and Going ...                              69

chapter 7        "Little Pratt" Does Big Things                                                    77

chapter 8        Great Engines and the Great Engine War                                            91

chapter 9        New Engines, New Challenges                                                       99

chapter 10       The Super Bowl of Fighter Engines and Big Twins                                  111

chapter 11       A New Company for the New Century                                                123


bibliography                                                                                      137

appendix A       Where Did the Eagle Come From?                                                   141

appendix B       Aircraft and Rocket Engines                                                      142

index                                                                                             159

# PREFACE

Pratt & Whitney's story of Dependable Engines is woven through the fabric of aviation history. Perhaps no other endeavor has so captured peoples' imagination as flight, and that is what inspired the company's founders back in 1925 when the revolutionary Wasp engine opened new horizons of speed and dependability. It is what inspires Pratt & Whitney people to this day.

Even with the glamorous aura that surrounds flight, the engine business has always been tough. The technology challenges have been, and remain, formidable. The risks have always been great. But what is most important is that Pratt & Whitney people—now more than 38,500 working in 22 countries around the globe—have always been smart, visionary, and absolutely dedicated to providing the best solution for our customers.

Their story is one of invention and reinvention. Leaders of Pratt & Whitney have had to reinvent their company repeatedly to meet new challenges and capture new opportunities—as they did during World War II and at the onset of the Jet Age. We are doing it again today, as Pratt & Whitney expands its engine service business with the skills and knowledge honed during almost nine decades as a leader in designing and building engines. We have become a global manufacturer with people and facilities located from Canada to China and from Norway to New Zealand.

Some common threads in Pratt & Whitney's history go back to the earliest days of company founder Frederick Rentschler and still apply today. He believed in assembling the most talented team he could, trusted colleagues with a passion for aviation and a commitment to excellence in engineering and manufacturing. The goal was, and remains, to ensure that customers are not just satisfied, but delighted.

As the history published by Pratt & Whitney on its 25th anniversary in 1950 said, "In aviation second-best tended to be no good at all." In 1925, Rentschler and his inner circle foresaw that aviation was growing beyond a one-man show and was destined to become an industry like automobiles or the railroads. They wanted a company with sound finances and a strong bottom line that allowed them to continue investing in new technology because that was the only way the fledgling industry could grow. Rentschler's belief that "the best airplane can only be designed around the best engine" has helped propel aircraft and engine technology to once unimaginable levels of power and performance. Over the years we've expanded that philosophy to include business jets, helicopters, military and commercial aircraft, even spacecraft and power stations!

Today, Pratt & Whitney people are justly proud of what they have accomplished as aero engines evolved from the piston-driven Wasp to the venerable JT8D to thrust vectoring, stealthy, supercruise engines for fifth-generation fighter aircraft. In the pioneering tradition of our founders, we will continue to provide innovative solutions to 21$^{st}$-century challenges—both in our air-breathing and liquid rocket propulsion engines and our unique customer service concepts. This includes tackling tough issues such as the need to dramatically improve fuel efficiency while drastically cutting noise and emissions. We are doing this today by employing a new step-change technology, the Geared Turbofan™ engine.

The engine business has undergone many changes since Pratt & Whitney first opened its doors, but one thing remains as true now as it was then, our commitment to the promise that founded our company—those two words under the Eagle: Dependable Engines.

Stephen N. Finger
President, Pratt & Whitney
June 2008

ix

# ACKNOWLEDGMENTS

The author offers his apologies in advance for any sins of commission or omission in this work. Not every story and every program is covered. I have relied on the best sources I could find as noted in the bibliography, as well as first-person interviews. The recollections of some people may be different from what I have related here but I have tried to be as accurate and fair as possible.

Every writer depends on the work, help, and guidance of others. I would like to thank Pratt's volunteer archivists, Jack Connors and Jesse Hendershot, who preserved the work of the late Harvey Lipponcott and Anne Millbrook. Also Priscilla Ubysz of Pratt's library services, Shannon Phelps, Jeanne Archambault, Marnie Walton, and Joanne Gagnon and David Pipins gave invaluable help. And a special thanks to Pratt & Whitney photographer Greg Roberts for help in choosing photos for this book. I would also acknowledge the support of Pratt & Whitney President Steve Finger.

Mark P. Sullivan
June 2008

# DEPENDABLE ENGINES
## THE STORY OF PRATT & WHITNEY

# Mr. Rentschler
## Comes to Hartford

## chapter 1

## ROOTS

**THE STORY OF PRATT & WHITNEY** is the sum total of thousands of stories of talented engineers, designers, machinists, sales and business people, office staff, secretaries, and support staff who built the company over almost 90 years.

Still, the company they built was principally the vision of one man, a quiet Ohioan named Frederick Brant Rentschler. A shy, reserved man, Rentschler was one of the first to see clearly that aviation could become a powerful industry, critical to defense and the economy of the United States. It is a mark of his vision that he founded the two most successful engine companies of the early days of aviation, Wright Aeronautical and Pratt & Whitney.

Rentschler's vision was built on three simple principles that guided him and are still part of the fabric of Pratt & Whitney today: a passion for aviation, a devotion to engineering and manufacturing excellence to satisfy customers, and the need for a sound business plan. Rentschler believed to the marrow of his bones that a company had to make healthy profits today so that it could continue to reinvest in the technology the industry needed to succeed years down the

Pratt & Whitney founder Frederick B. Rentschler with his own Vought 02U Corsair.

road. Author and engine historian Graham White may have described Rentschler best: "He was one of the few men gifted with the ability to operate machine tools, understand engines and yet have the business acumen that is so key to success."

We should note that Rentschler's story is no Horatio Alger tale. He came from a successful Hamilton, Ohio, family that owned its own foundry and manufacturing company. The Rentschler family had strong business connections throughout the country, which he would use. For instance, it certainly did Frederick Rentschler's dream no harm that his brother, Gordon, would rise to become chairman of National City Bank, today's Citicorp.

Rentschler graduated from Princeton in 1909 and went to work in the family business in Hamilton. His first foray into aviation came with the U.S. entry into World War I in 1917. Commissioned as a lieutenant, he was assigned to inspect and supervise aircraft engine production in the New York region.

Rentschler believed in careful planning, focus, and order, things he saw little of in America's efforts to "darken the skies of Europe" with military airplanes. He saw not an industry but a group of small disjointed enterprises "hopelessly behind that of any of the great powers."[1] The massive efforts to turn the automotive industry into airplane and engine manufacturers with their own designs resulted only in a trickle of semi-obsolete planes and engines getting to Europe. To Rentschler's orderly way of looking at things, "the whole venture proved completely ineffective from the standpoint of the war." His own work in the Army had involved inspection at the Wright-Martin plant in New Brunswick, New Jersey, in building proven and up-to-date French Hispano-Suiza engines, a path he felt the country should have followed more generally. He summed up in his usual analytical way: "Our knowledge of aviation was so hopeless and the time so short that this [building proven European engines] at least might have furnished a way of effective help, but pride required us to follow the other process."

One might think that this experience would have soured the young man from Hamilton on aviation, and Rentschler himself said that he expected to return to the family business after the war. That was certainly something his father, George, expected. He called aviation "a damn fool business, mostly for sportsmen." But his son had been bitten by the bug like so many before and after him. After the war the opportunity came to run a newly organized engine company, Wright Aeronautical, which had emerged more or less from the Wright-Martin operation Rentschler had known.

---

[1] All direct quotations of Rentschler, unless otherwise noted, are taken from his 1950 reminiscence, "An Account of the Pratt & Whitney Aircraft Company" which was published privately.

From 1919 to 1924, Rentschler ran Wright successfully and could be credited with laying the groundwork for its most famous engine, the Whirlwind, which powered Charles Lindbergh across the Atlantic in 1927. That engine evolved from the work of radial pioneer Sam Heron on cylinder design and Charles Lawrance's Lawrance J engine. Lawrance did not have the wherewithal to actually produce his design, so, under pressure from the Navy, Rentschler acquired Lawrance's company for Wright, a decision he never felt comfortable with and which led in part to his departure from Wright. Heron would later join Wright briefly and work on the final design of the Whirlwind. His work would also greatly influence the design of Pratt & Whitney's radial engines.

It was at Wright that Rentschler first put together the men who would also become the nucleus of Pratt & Whitney. George Mead, a 1915 Massachusetts Institute of Technology (MIT) graduate, had been an experimental engineer at Wright-Martin and also worked at the Army research facilities at McCook Field. He was a remarkable planner and understood the reality of manufacturing requirements as well as the theory of engineering powerplants. Don Brown was a sound factory manager and businessman. Andy Willgoos was a designer who "could think with his fingertips." People remarked that he simply had a "feel" for the right design of a part. Charlie Marks was an expert on acquiring the tooling needed for the job. Jack Borrup was a crusty shop guy and went to work as shop superintendent. Mead, Willgoos, Marks, and Borrup had also served under the legendary Henry Crane of the Crane-Simplex company in their early careers. Crane was close to Rentschler and actually served as the temporary vice president of engineering at Wright Aeronautical.

Wright was doing well competing with Curtiss and an emerging Packard. "I think it would be fair to rate it as the outstanding engine company of the period from an all-around viewpoint," Rentschler later recalled. But he was not a happy executive. He had achieved the goal of creating a successful engine company that was a sound business, but was having trouble with the idea of plowing profits back into Wright. His board of directors, Rentschler said, was "almost all investment bankers, none of whom had any real appreciation of what we were trying to do." So in September 1924, Rentschler quit, not sure exactly what he would do next.

> "I realized then and there that I would never be happy unless I could find some way of going on with aviation."

He chewed it over for a few months and decided he was not going back to Hamilton and the family business: "I realized then and there that I would never be happy unless I could find some way of going on with aviation."

As he contemplated exactly how to do that, his vision crystallized around one central idea: "It seemed very definite that the best airplane could only be designed around the best engine."[2]

# BEGINNINGS

There were opportunities in the fledgling aviation world. Historians love to make connections, and one could draw a line from the naval race between Britain and Germany in World War I to the Washington Naval Limitations Treaty of 1921 to the opportunity Fred Rentschler saw on the horizon. That naval treaty specified limits for capital ship tonnage for the world's major naval powers. At the time the U.S. Navy had two huge battle cruisers under construction, Saratoga and Lexington. They could not be completed as big gun ships but could be converted into that new type of warship, the aircraft carrier. So by 1925 the U.S. Navy had what were then the two largest aircraft carriers in the world under construction and needed some 200 new aircraft to outfit them.

> **❝ It seemed very definite that the best airplane could only be designed around the best engine. ❞**

The Navy had pushed Wright into air-cooled radial programs but was worried about being dependent on just one company. In early 1925 Rentschler paid a visit to Admiral William Moffett, head of the Navy's Bureau of Aeronautics and a tough infighter in the battle for funding for aviation in the post-World War I defense cutbacks. Rentschler said he thought he could put a team together and create a lightweight, air-cooled, radial engine of 400-plus horsepower that would be just the ticket for those 200 new planes for the Saratoga and Lexington. Moffett and the head of the bureau's engine section, Commander Eugene Wilson, strongly encouraged Rentschler, whom they knew well from Wright. But Moffett made it clear there was no development money available. If Rentschler could do it, the Navy would be interested, and there might even be money available for some experimental engines if the first one worked out.[3]

The Rentschler family connections then came into play. Brother Gordon at First National City knew that the Niles-Bement-Pond Company was looking to invest some of its excess cash in new ventures for its Pratt & Whitney Machine Tool Company in Hartford, Connecticut.[4] Connecticut was an ideal location for setting up a company that needed precision manufacturing. The American machine tool

---

[2] Rentschler is often quoted as saying, "The engine is at the heart of the matter." Actually he did not coin that phrase. Rentschler himself credited it to the company's first public relations (PR) man, George Wheat.

[3] Wilson would later join Rentschler at United Aircraft Corporation, where he rose to president and vice-chairman.

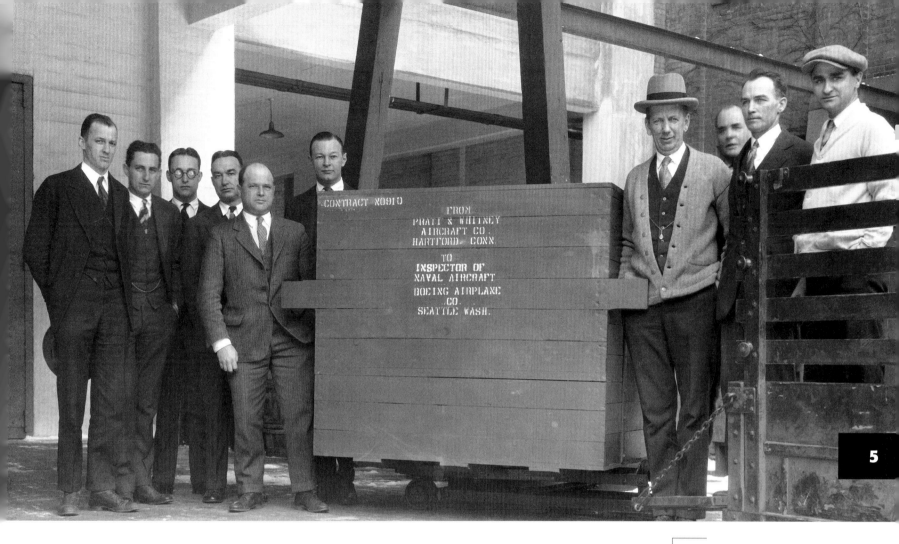

CONTRACT N0910
FROM
PRATT & WHITNEY
AIRCRAFT CO.
HARTFORD, CONN.
TO
INSPECTOR OF
NAVAL AIRCRAFT
BOEING AIRPLANE
CO.
SEATTLE, WASH.

industry had been virtually born in the region going back to Eli Whitney and the Springfield Armory. Some of the most skilled metal workers in the world populated the region, and still do.

It probably was not a coincidence that the chairman of Niles-Bement-Pond was James Cullen, an old friend of George Rentschler. Cullen had at one time operated Niles Manufacturing Company next door to the Rentschler plant in Hamilton. So Frederick Rentschler had no trouble arranging a meeting and outlining his plans for a new aircraft engine company. He would need an initial investment of $250,000 (with a possible $1 million after that), machine tools, and factory space. Cullen sent his old friend's son up to Hartford to see the folks at Pratt & Whitney. They liked Rentschler's ideas and showed him some space he could move right

The first Wasp is shipped to the Navy.

---

4 Pratt & Whitney Aircraft was twice removed from the original Pratt & Whitney Company. Francis Pratt and Amos Whitney started out as mechanics in Sam Colt's pistol factory in Hartford. They formed their own company in 1860. It was famous for the measurement accuracy of its machine tools. They basically created the commercial standard inch measurement, which had not existed previously. Both were long dead by the time Pratt & Whitney Aircraft got underway.

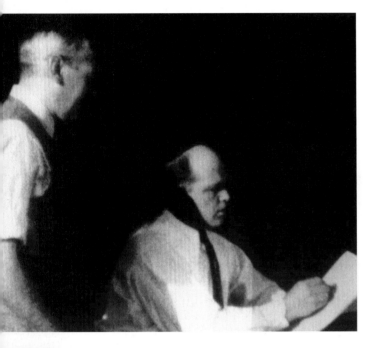

Erle Ryder and Andy Wilgoos at work in the New Jersey garage.

into. It was the old Pope-Hartford auto plant on Capitol Avenue, presently filled to the rafters with Connecticut River Valley cigar tobacco.[5]

As the business negotiations proceeded, Rentschler was contacting his old associates at Wright-Martin. When he left in 1924, Rentschler had promised he would let them know if he ever did anything again in aviation. Apparently, they were as uncomfortable with the Wright investment banker board as Rentschler had been. Because of Rentschler's quiet leadership and passion for aviation, George Mead, Don Brown, Jack Borrup, Charlie Marks, and Andy Willgoos all agreed to quit Wright and go to Hartford.[6]

On July 14, 1925, the contract that created the Pratt & Whitney Aircraft Company was signed.[7] Rentschler and Mead would own half the stock. Pratt & Whitney Machine Tool would own the other half along with preferred stock for the capital it had advanced the new engine company. One thing Rentschler insisted on was that the new company be completely independent of Niles-Bement-Pond and Pratt & Whitney Machine Tool, although Niles-Bement-Pond got two seats on the six-man board. Rentschler did not want to repeat his experience at Wright-Martin.

Even before the facilities in Hartford could be emptied of tobacco, spruced up, and outfitted to build engines, Mead and Willgoos had started working on the preliminary design of a new engine for the Navy in Willgoos' garage in Montclair, New Jersey. The parameters were basically simple. The engine had to produce at least 400 horsepower and could weigh no more than 650 pounds. At the time no one had built anything close to that.

The fledgling company was jumping feet first into one of the hottest debates in aviation at the time: air-cooled vs. liquid-cooled engines. The backers of liquid-cooled engines believed that those powerplants could develop more

---

[5] Company legend says that when the first factory people moved in there were still bales of tobacco about and workers, who could not smoke on the shop floor, would reach over and grab a chaw. The author, who grew up on a Connecticut valley tobacco farm, can attest that this is something you would only do once after you tasted raw cigar wrapper tobacco.

[6] It is said that no one except George Mead and Don Brown ever called Rentschler "Fred." To everyone else he was "Mr. Rentschler."

[7] Pratt & Whitney never actually built an aircraft, but the name still sticks. For many people in Connecticut, Pratt & Whitney is still "The Aircraft," and the people working there are "aircrafters" or "Pratt Rats."

horsepower than air-cooled radials and had much less of a problem with drag because they were more streamlined. Air-cooled advocates, the new Pratt team among them, argued that the latest engineering offered the ability to build a powerful, yet lightweight, air-cooled radial. If properly designed and installed, its larger frontal area would offer no more "head resistance" than a liquid-cooled engine and would have all of the cooling necessary for higher horsepower. A team could also build a smaller, lighter, and faster aircraft with a radial than a liquid-cooled engine. And in Navy carrier launch/landing or even catapult operations, an air-cooled engine would be much more robust than a liquid-cooled with complex and often fragile radiators, pipes, and hoses.

Lt. Commander Bruce Leighton, Wilson's predecessor in the aeronautics bureau's engine section, called the air-cooled radial the Navy's great hope:

The Wasp on the test stand at the Capitol Avenue plant.

> There is less sense in liquid-cooling an aircraft engine than in air-cooling a submarine. . . . Each of those plumbing pounds requires another pound of wings and tail to lug it around. But with the air-cooled engine we can throw away the plumbing and convert that dead weight into payload with a smaller airplane. On board ship you've got to keep 'em small or leave 'em off.

Pratt & Whitney always stuck with air-cooled radials, believing they offered the best path to higher horsepower and reliability. There was one notable exception, much to the company's regret, as we shall see a bit further on.

# GETTING UNDERWAY

As the factory was being readied in July, Mead and Willgoos were working away in that garage in Montclair and needed some extra help. Thus, Pratt & Whitney got what was arguably its first employee. None of the original team would see a dime until they got to Hartford. But Mead and Willgoos needed an assistant, and so Willgoos hired an old friend, Erle Ryder. Rentschler paid his salary out of his own pocket, but Erle Ryder was the first person to draw a pay packet from Pratt & Whitney Aircraft.

By August, things were ready in Hartford, and Rentschler and his team packed up their families and moved, as he put it, "down east." Late in his life he recalled matter of factly, but with a tinge of nostalgia:

> There were no Chamber of Commerce or Rotarian representatives to meet us, and I can truthfully say we didn't even cause a ripple as we quietly stowed ourselves away in our new homes in Hartford. In fairness to our new community, it is perfectly honest to point out that there was just no reason in the world why they should have paid us any attention. The morning of August 3rd found us all on the job.

There were three offices, one each for Rentschler, Mead, and Don Brown as factory manager. A fourth office was the drafting room. There Andy Willgoos led the designers, himself, and two others. Jack Borrup and Charlie Marks scoured the machine tool company for equipment and the people who could build the first engine. Mead released drawings to Brown, who was also purchasing agent because he really had not much factory to manage. He in turn passed them through a small hole in his office wall to Borrup, who sat atop a small platform on what was the "factory floor."[8]

The design that Mead and Willgoos came up with was a breakthrough for its time. In those days radial engines hit the redline at about 1800 rpm. The new design used a split rather than solid crankshaft, forged instead of cast crankcase, and a solid master rod that cut weight allowed for 1900 rpm in normal operation and up to 2400 rpm in a dive with higher pressures in the cylinders. They aimed at a nominal rating of 420 horsepower on 1340 cubic-inch displacement to ensure meeting the Navy's 400-horsepower requirement. The design because it was lighter allowed for the bigger displacement of 1340 cubic inches compared with its closest competitor, the Wright Simoon design at 1140 cubic inches, which could run at just 1650 rpm. Mead and Willgoos also worked out a scheme to machine cooling fins from a solid so they could be thinner and could get more on each cylinder.

By Christmas Eve 1925 the first engine was assembled, and board member Edward Deeds made good on his promise of a holiday turkey for everyone if the engine was done by Christmas. So all 20 or so employees put in their orders, and Mrs. Rentschler and Mrs. Mead handed out the food baskets.

The engine was built, but what to call it? Later Rentschler recalled: "Dozens and dozens of suggestions were thrown back and forth within our little group. Finally we began gravitating toward "Bees" as a general designation for our engine types and, according to my best recollection, my wife (Faye) suggested Wasp for the name of our first product."

---

[8] The first equipment consisted of two lathes; two milling machines; a boring mill; two grinders; a vertical, a horizontal, and a hand-milling machine; a radial drill; two small drills; and some benches with vises.

The engine had a name, and it weighed almost exactly 650 pounds, but would it meet performance specifications? Heck, would it even start? It did, and by the third test run achieved 425 horsepower.

# SUCCESS

Frederick Rentschler was not a man given to hyperbole, but one can almost see the grin in his memory of that day: "It (the Wasp) ran as clean as a hound's tooth and was actually just the thoroughbred that it looked. These characteristics had never been previously achieved in an aviation engine."

Navy tests began in March 1926, and the Wasp breezed through them.[9] The first flight was on May 5, 1926, when Lieutenant C. C. Champion took off from a reclaimed mud flat on the Anacostia River in Washington, D.C., in a Wright Apache powered by the second experimental engine. The Navy then installed Wasps in the Boeing F2B, the Curtiss Hawk, and the Vought O2U-I Corsair. In each was Rentschler's thoroughbred. In October, the new company got its first production contract for those 200 Navy engines, and by February 1927 the plant was turning out 15 a month with the production numbers going up steadily. Another new engine was well underway at Pratt & Whitney, the 525-horsepower Hornet, underlining Rentschler's belief in constant investment and product improvement.

The acid test for both engines came in 1929 when the new carriers Saratoga and Lexington set sail on full-scale maneuvers with 200 brand new Wasp- and Hornet-powered aircraft. Over three months, they flew 2100 missions with 6000 hours of completely trouble-free engine operation. According to company lore, the chaplain of the Saratoga gave thanks for the "dependable engines," and that is how the

9

The Wasp would prove itself onboard the USS Saratoga and her sister ship, the USS Lexington.

---

[9] The first Wasp never flew. After initial testing, it was donated to the Franklin Institute in Philadelphia for exhibition. It is now in the collection of the National Air and Space Museum.

famous Pratt motto came about. It is a good story, but the motto had been used on the company's eagle trademark before 1929. Actually, the chaplain had only used "dependable engines" as a text for a sermon. There is no existing record of how "Dependable Engines" came about except that in the 1920s dependable engines was an important phrase to convince people that aviation was more than just a fad for the very brave or the very foolhardy.

Charting a course Pratt & Whitney follows to this day, both were constantly improved. The R-1340 Wasp eventually reached 600 horsepower and the R-1690/R-1860 Hornet 875 horsepower. Between 1925 and 1960 Pratt built nearly 35,000 Wasps and between 1926 and 1942 just under 3000 Hornets. Combined they powered some 160 different aircraft types. And by 1931 the company was working on twin-row radials to get more power for faster speed and more payload and range while not increasing drag with a larger and larger frontal area. The

> " According to company lore, the chaplain of the Saratoga gave thanks for the 'dependable engines,' and that is how the famous Pratt motto came about. "

14-cylinder Twin Wasp had 1350 horsepower compared to the ratings of the Wasp and Hornet. In 20 years of production, an astounding 173,618 Twin Wasps would be built at 1830 and 2000 cubic-inch displacement. They would power some 80 aircraft types.

Today aerospace is the global industry Rentschler foresaw, with multi-billion dollar development programs. It is interesting to contrast this with those days in 1925. Six engineers and some 20 shop people designed and built the three original Wasp prototypes and the first Hornet in nine months at a cost of $202,713.29.

# ON THE MOVE

By the late 1920s the size of Pratt & Whitney's facilities was very much an ongoing concern. The company was rapidly outgrowing the old Pope-Hartford plant and needed room to expand production. So Rentschler and company packed up again, only not so far this time, just across the Connecticut River. An 1100-acre parcel of land, a good deal of it tobacco fields, was purchased, and in 1929 Pratt moved to its present location on Main Street in the south end of East Hartford.[10]

There was another reason for the move. Rentschler and his team had worked with virtually all of the early aircraft companies of that time. He saw an oppor-

---

[10] The plant site might be the only aerospace factory that contains the grave of a Revolutionary War soldier. Sgt. Herman Baker of Tolland, Connecticut, died in East Hartford while traveling home on January 21, 1777. He had been held after the British captured him in September 1776. He was released when he contracted smallpox.

tunity to realize his vision of a thriving, dynamic aviation industry by combining forces. So space was needed for a brand new corporate headquarters for something called the United Aircraft and Transport Corporation. His own words best describe those times:

> We also began to realize that to continue to keep pace with the fast growing aviation business we needed to work very closely with allied companies, particularly in plane manufacturing and transport. My close personal relations with (William) Boeing and (Chance) Vought made it natural first that we work together as we had proved we could do so very successfully. From all this there naturally emerged the possibility of joining together in some more orderly and effective fashion. . . . Inevitably Boeing, Vought and I began discussions of the possibility of a merger of our interests.

The United Aircraft and Transport Corporation (UA&TC) that the three aviation pioneers put together was something to behold. Wags at the time referred to it as "The General Motors of the Sky." UA&TC consisted of Boeing Airplane Company, Boeing Aircraft of Canada, Hamilton Standard Propeller Corporation, Northrop Aircraft, Pratt & Whitney Aircraft, Pratt & Whitney of Canada, Sikorsky Aviation Corporation, Stearman Aircraft, and Chance Vought Corporation. There were also three transport companies in the fledgling airline business: Boeing Air Transport, Pacific Air Transport, and Stout Air Services. Also the new corporation held United Airports of Connecticut and United Airports of California.[11]

William Boeing was chairman, Rentschler president, and Vought vice president of the new corporation. The technical committee included among others Mead, Claire Egtvedt of Boeing, Thomas Hamilton of Hamilton Aero, Frank Caldwell of Standard Steel Propeller, John Northrop, Leonard (Luke) Hobbs of Pratt & Whitney, and Igor Sikorsky.[12]

The first annual report in 1929 proudly proclaimed that UA&TC "occupies a unique and possibly the strongest position in the aeronautical field of any company in the world." And as Wall Street might say today, they had "good numbers." Sales in 1929 were $31.5 million. Profit was $8.3 million. The corporation had $27.4 million in assets and just under $16 million in cash.

---

[11] The airport companies are little known today. The Connecticut operation built Rentschler Field adjacent to the headquarters in East Hartford. Closed in the 1990s, it is now the home of the University of Connecticut football stadium, still called Rentschler Field or "The Rent." United Airports of California built what is now the Burbank Air Terminal, proudly proclaiming in 1929 that it was "only 15 minutes from Hollywood."

[12] Hamilton Aero and Standard Steel were merged at this time. The story is that Thomas Hamilton and Harry Kraeling of Standard Steel did not get along, so Hamilton was sent to Los Angeles as a salesman for the burgeoning aircraft companies around Los Angles and to supervise United Airports of California. He became a dashing figure on the Los Angeles–Hollywood scene and is often described as the most colorful character among the early executives. He later played a prominent and sometimes controversial role as United's sales representative in Europe on the eve of World War II.

All three of Rentschler's principles—a passion for aviation, technical excellence, and sound business practices—are evident in that 1929 annual report. Each of United's companies "has a record of successful performance and has contributed in a distinguished manner to the progress of aviation. The strength of United's position is due in substantial measure to the unmatched engineering staffs of its subsidiary companies."

By 1930, Pratt & Whitney had grown to about 1000 employees. Like all companies, Pratt has its stories and legends of what it was like in the good old days. In 1926, a young Yale engineering graduate, H. Mansfield Horner, joined the company as a "stock chaser." Of course, inevitably, he would become "Jack" Horner, and, perhaps not so inevitably, he would become head of Pratt & Whitney and eventually chairman of United Aircraft. After his retirement, he recalled running from supplier shop to supplier shop to get a couple of studs plated or pick up a single crankshaft. In the late 1920s, he served as paymaster, and that included the nightshift. The company paid in cash with most people making about 30 cents an hour. Horner recalled that he was worried delivering the pay envelopes at midnight:

> I complained to my boss that somebody is going to knock me in the head
> one of these times and nobody will find me until the next morning. So for
> protection he gave me a .32 automatic which he said don't ever shoot, but
> show it to everybody so they know you have it and that will protect you.
> Well it did. Nobody ever bothered me.

Many years later it took armored cars to deliver the payroll, still in cash, to the East Hartford plant. Streets would be closed, and Pratt security guards and police were on duty with pistols and shotguns.

Another story involves a somewhat elderly janitor inherited from Pratt & Whitney Machine Tool. Not too long after his famous transatlantic flight, Charles Lindbergh paid a visit. The Lone Eagle showed up an hour early and wandered into the plant. When the welcoming delegation finally found him, he was being escorted by the janitor who was pointing out the wonders of the shop with his broom handle.

Carroll Brooks was an accountant who had come to Pratt from Hartford's very Yankee conservative insurance industry. Having turned out in his best suit and tie for his first day of work, Brooks recalled that he was immediately told "have your coat off when you come in to go to work in the morning." After the staid atmosphere of an insurance company, Pratt & Whitney was sure different. "Here were all the executives walking around in shirtsleeves, talking in the saltiest

The Capitol Avenue
production line
begins to hum.

language you ever heard and gathering together every once in a while to dash out into the shop like a football team." Taken aback at first, Brooks conceded that quickly, "I got caught up with the dash of the whole thing."

Things changed slowly over the years. One Pratt engineer recalled coming to work straight out of college in the late 1960s decked out in his colorful double-breasted blazer, wide tie, bell bottoms, and platform shoes. "Go home and change. We don't wear our pajamas to work at Pratt & Whitney," he was told by a stern-faced boss.

Many stories abound about shop superintendent Jack Borrup and his high-pitched lectures, often punctuated with strong Anglo-Saxon epithets. A memo supposedly came out ordering changes in the shop without Borrup's okay. When he saw the memo, he picked up his phone, dialed the man who had signed it, said only "I still work here," and slammed down the phone. In a company history, machinist John Aronson recalled getting a Borrup tongue-lashing and then having Borrup put his hand around Aronson's shoulder. "John, you're doing fine," the superintendent consoled, "That was just a little lesson for you. In this business you have to have a hide like an elephant."

So as the 1930s opened, it appeared that Frederick Rentschler and his colleagues were well on the way to their vision of a world-class aviation company. The 1930s, however, would prove a difficult time for them, their company, and the industry they had helped start in a few short years.

# Clouds
## on the Horizon

## chapter 2

**THE 1930S WOULD PROVE TO BE** one of the most difficult times in the history of Pratt & Whitney, even as its technical success with the Wasp and Hornet continued. Like every other company, Pratt was hit by the Depression as well as parsimonious defense budgets. It also became a target of some of the reforms of the New Deal, and the close-knit group of company founders began to fray. And despite the outstanding success of the Wasp and Hornet, strong competition emerged from Wright and Allison.

## AERIAL HIGHWAY

Although Pratt & Whitey's initial success came from the military, it was not long before commercial air transport began to play a large role. The Pratt & Whitney team had always built their plans around bigger, powerful engines. They had not bought into the idea that every American would have a light airplane parked next to the family *flivver*. So it turned out that the timing of the Wasp and Hornet would be perfect.

Several factors were combining in the mid-1920s to get air transport out of the cow pasture and beyond the barnstormers in their Liberty-powered Jennies. The Morrow Board and the Congressional Lampert Committee had concluded that the U.S. aeronautical

William Boeing and Frederick Rentschler inspect the first Wasp for Boeing's airmail operation.

The Wasp and its big brother, the Hornet.

industry was weak and ill-founded and that almost every European nation was miles ahead of the United States. Their recommendations boiled down to the need for a sustained, long-term military aircraft program, and that to do this the government had to foster a financially strong and technically competent aviation industry. The government should get out of the business of building airplanes itself and end the practice of auctioning off production contracts to the lowest bidder after a company had developed a new airplane or engine. These findings were followed by concrete legislation such as the Air Mail Act of 1925 and the Air Commerce Act of 1926, which could be said to have laid the foundation of American commercial aviation.

Airmail was a critical issue because in the mid-to-late 1920s no one had figured out yet how to build an air transport business just carrying passengers. It was airmail that brought William Boeing and Frederick Rentschler together and could be said to be the linchpin in building UA&TC ( United Aircraft & Transport Corporation).

The Post Office Department's United States Air Mail Service had pioneered air transport in the early 1920s. Charles Lindbergh began his career there. By 1925 there was enough of a system for Congress to authorize turning over airmail service to private contractors. In 1928 Boeing Air Transport won the Chicago to San Francisco airmail route, the longest in the country, with a bid of $2.89 per pound. The nearest competitor had come in at $5.09 a pound. The competitors smelled a rat and accused Boeing of underpricing to ensure a market for his airplanes and to build an airmail monopoly. The competitors were looking at things from the point of view of the standard mail plane of the day, the DH-4. It could carry only 500 pounds of mail under ideal conditions and usually could haul only between 250 and 350 pounds. Boeing had designed a sleek new mail·plane, the 40-A (later the 40-B), and William Boeing had heard great things about the new Wasp engine. The 40-A/B with a Wasp as opposed to a Liberty could carry 1500 pounds of mail and two passengers for an additional $400 in revenue per trip. The numbers worked, and Boeing Air Transport was a success.

> **It was airmail that brought William Boeing and Frederick Rentschler together and could be said to be the linchpin in building UA&TC.**

When Boeing, Rentschler, and Vought formed UA&TC, air transport was a key element, and they quickly added Pacific Air Transport, Varney Airlines, National

Air Transport, and Stout Airlines. These were all brought together under the aegis of a new subsidiary, United Airlines.

The success of the Boeing mail plane led other manufacturers to look to the Wasp and Hornet to build commercial transports. The Ford Tri-Motor, the Fokker single-engine and trimotors, Igor Sikorsky's flying boats, the early designs of Allen and Malcolm Loughead and Donald Douglas all used Pratt power. The long relationship between Juan Trippe's fledging Pan American and Pratt & Whitney led by the late 1930s to the clipper services spanning the Atlantic and Pacific, one of the most romantic eras in commercial aviation.[1]

# DAYS AT THE RACES

Speaking of a romantic era, the 1930s also saw the high point of air racing. The Thompson Trophy, the National Air Races in Cleveland, the Bendix, the Shell speed dashes, the Granville brothers and their Gee-Bee, Jimmy Doolittle, Jacqueline Cochrane, and Roscoe Turner and his lion Gilmore all evoke memories of those days. And there was Wiley Post, Amelia Earhart, and Howard Hughes setting speed, endurance, and distance records. Pratt-powered racers won eight out of nine prewar Bendix races and eight out of 11 Thompson Trophies. And when the last Thompson winner taxied in in 1949, it was Cook Cleland's surplus Goodyear Corsair powered by a massive Wasp Major.

Roscoe Turner, perhaps the most colorful of the 1930s air racers.

17

There are dozens of wonderful stories about those racing days. One year at the Thompson, Roscoe Turner was so broke he had to borrow a clean shirt from a Pratt representative so that Turner could attend the dinner the night before the race, and Pratt paid for his seat. Well, Turner won and threw a champagne and caviar party for the Pratt team and anyone else he could find, blowing half his prize money on one night.

---

[1] Pan Am's technical advisor was Charles Lindbergh who worked closely with Pratt engineers. The desk he used is still at the company. In 1990 Pratt celebrated its 65th anniversary with a huge air show at Rentschler Field. Company employees burst into applause when a Pan Am A310 arrived with "'Clipper Pratt & Whitney" emblazoned on its nose.

H. M. "Jack" Horner, who would eventually become chairman of United Aircraft, recalled that Pratt would often stake racing pilots to an engine and support for the publicity value and hope to get some money back if the pilot won anything. Even Howard Hughes got an engine for a transcontinental speed record try in the mid-1930s. In an interview after his retirement, Horner said he did not realize how much money Hughes was actually worth and got a little irked when the billionaire neglected to pay for his engine:

> We'd see in the paper where he was out on some private yacht cruising around Block Island Sound with some lady friend. I was going tuna fishing out there that weekend. We happened to come across this yacht. I got the bull horn out of the fishing boat and started blasting, "Howard, where the hell is that check." He never showed on deck and I don't know whether he got the word or not. But eventually we did get paid, believe it or not.

# STUMBLES

In the early 1930s the success of the late 1920s seemed to be continuing, Depression or not. The military business was on sound footing, albeit with small budgets. The air transport business was taking off, and racing was gaining Pratt & Whitney a great name. But the clouds were thickening on Pratt's horizon.

Boeing had designed what was arguably the first truly modern airliner in 1930, the 247, a low-wing, all-metal streamlined twin-engine ship with retractable landing gear. Naturally it would have a Pratt engine, keeping things neatly all in the UA&TC family. But which engine? The Pratt engineers wanted to use the Hornet. The plane could then have a gross weight of 16,000 pounds. But the United Airlines pilots said they would not fly the thing. They loved their Wasps and said a 16,000-pound airplane was too heavy for the airports of the day. Some in the Boeing engineering community were also skeptical about a larger, heavier design. So the 247 was scaled back to 12,000 pounds with a 12-passenger capacity. It went into service in 1933

Howard Hughes, whom Jack Horner dunned for a bill.

The revolutionary Boeing 247 airliner.

with Wasps.

Down in Los Angeles Donald Douglas saw an opportunity. Using Wright Cyclones, which had about the same power as the Hornets envisioned for the 247, Douglas built first the DC-2 and then the remarkable DC-3. And he had an open market. Because Boeing and United Airlines were all part of UA&TC, no other airline had a shot at getting 247s until United Airlines' order was completed. They flocked to Douglas.

The UA&TC juggernaut was losing some of its momentum and worse was to come. The episode is sometimes called the Air Mail Scandal. By 1934 the Depression was at perhaps its deepest point, and the country had come to bitterly resent the financiers and power brokers people blamed for the crash. There was also a current running against arms manufacturers, perhaps left over from the "merchants of death" days following World War I. Franklin Roosevelt had brought the New Deal to Washington. In the Senate, Hugo Black, later an associate justice of the Supreme Court, started hearings on just how airmail contracts had been handed out and how much money the companies and their executives had made. Boeing, Rentschler, Mead, and other UA&TC executives were grilled unmercifully about the profits their companies had made and their own considerable personal fortunes.

> **"The air transport business was taking off, and racing was gaining Pratt & Whitney a great name."**

Advocates said the airmail system had been set up partially to foster the American aviation industry and contracts had been allocated only to companies that were financially robust. In 1930 Postmaster Walter Brown had awarded UA&TC and a couple of other firms almost all of the mail contracts because he said they were sound operations. Brown told the Black committee that at the meeting he had "exerted every proper influence to consolidate the short, detached and failing lines into well-managed, well-financed systems." Critics called the session "The Spoils Conference," and it just stuck in the nation's craw in 1934.

President Roosevelt annulled all mail contracts in February 1934 and ordered the Army Air Corps to fly the mail. Army pilots and aircraft were just not suitable for the job, and five were killed in crashes almost immediately. The Roosevelt Administration backed off but proposed new legislation that by the end of 1934 would ban any company from an airmail contract if it had any connection with a company that made aircraft equipment. It meant the end of UA&TC. It should be noted that at

19

no time in all of the hearings and legislation was it ever found that the company or any individuals had broken any laws, even if the whole affair had a bad odor.

Rentschler later said of his own compensation package from UA&T, "Particularly in the late 1920s this was no unusual arrangement. Had we failed entirely, or if we had been unable to earn profits, our extra compensation provision would have meant exactly nothing."

William Boeing was so angry that he basically walked away from aviation entirely. Rentschler was also angry and embarrassed and pulled back from day-to-day management for a time. But he stuck with what became United Aircraft Corporation (UAC), made up of Pratt & Whitney, Hamilton Standard, and Vought/Sikorsky. In the 1970s, under the leadership of Harry Gray, United Aircraft became United Technologies and expanded greatly.

Rentschler summed up this bitter experience in his privately published 1950 recollections:

> Naturally the success of Pratt & Whitney and United made us a perfect target for the New Dealers. . . . There was no use in picking to pieces the companies which had gone bankrupt and had contributed nothing to aviation, so it was inevitable that they turned their main fire on United, whose only crime was that we had earned a reasonable profit in a field where most others had lost their shirts.

George Mead, the brilliant engineering leader of Pratt & Whitney.

# TROUBLE AND MORE TROUBLE

Having weathered the airmail storm, Rentschler probably hoped for more tranquil times ahead. It was not to be.

First, he saw that Pratt had become complacent. Wright and its Cyclone family had made a big come back with "a well thought out and well rounded off forward engineering program." Then the Army had designated the liquid-cooled Allison engines for all new fighters. The Army had even instructed Pratt to begin development of a liquid-cooled engine, something it had no experience in at all.

To the ordered mind of Frederick Rentschler, "Pratt & Whitney engineering was using a shotgun to say the least." During the 1930s, there were seven engine development programs, not counting the Army's liquid-cooled program or continued development work on the Wasp and Wasp Junior, a smaller engine for light planes and racers.

One cause of all this was the bad relationship between Don Brown and George Mead, or more correctly the lack of any relationship at all. When Rentschler became president of the old UA&T, he felt he had to name a president of Pratt & Whitney to watch out for day-to-day operations. The choices were Mead or Brown. Mead said he did not want the job. He liked Brown personally, but would not work for him. Rentschler felt he had no choice but to give the job to Brown. After the 1934 breakup, Brown also became United Aircraft (UAC) president. The relationship between Brown and Mead deteriorated quickly. Engineering programs got out of hand, and Mead was more and more detached, often working from an engineering office he had set up in his home, frantically trying to make a sleeve-valve, liquid-cooled design work.

In 1935 some progress was made when Leonard "Luke" Hobbs became engineering manager and started to sort out programs, although Mead was still company vice president. Rentschler called 1937 the "all-time low" for the company.

The new plant in East Hartford in 1930.

21

"Mead's general relationship with Brown and others was so involved that he was practically sidetracked." He would resign in 1939, suffering from total exhaustion.

Also, the books did not look as good as they once did. Pratt and Vought/Sikorsky were not doing well. In fact, those beautiful Sikorsky clippers had never made a dime. UAC was bailed out by Hamilton Standard (HSD). This held the patents on the variable pitch propellers that everyone needed. HSD kicked in 51 percent of UAC's total profit of $3.8 million in 1937.

There were, however, some encouraging signs. The French were beginning to order Pratt & Whitney engines with the rise of the Nazi military threat, and the British were soon to follow. Luke Hobbs had made great strides in straightening out engineering.

The company was concentrating on the development of the R-1830 Twin Wasp, the R-2000 Twin Wasp, and the R-2800 Double Wasp, all two-row air-cooled radials that could compete with the Wright Cyclones.

But there was another blow coming. Don Brown was ill, very ill. By early 1939 it was becoming clear that Brown was terminally ill with cancer. It perhaps says something of the fear that cancer brought into people's minds in those days that Brown apparently was never told the nature of his illness and that Rentschler was the one who got the word from his doctors. Don Brown, the man closest to Rentschler, died in January 1940.[2]

Rentschler named Jack Horner, who had started with the company in 1926, to run Pratt & Whitney and William Gwinn, who started not long after Horner, as the new assistant general manager. Both would rise to become chairmen of United Aircraft. Horner succeeded Rentschler, and Gwinn succeeded Horner.

> 66 Building aircraft engines is not a business for the faint of heart. 99

Of course, by now it was clear that war was coming and there was one more task to be accomplished, that liquid-cooled engine for the Army. In the spring of 1940, General Hap Arnold came up to East Hartford. He and Rentschler had known each other for years, and Rentschler buttonholed him on the engine. No matter how good it might be, there was no way it could be ready to help win the war everyone knew was coming. The three twin-row radials were doing well, and the company could build a 3000-horsepower radial if it could get out from under the liquid-cooled project that had so bedeviled Mead. Arnold gave the go-ahead, and the liquid-cooled project ended. Work began on what became the 28-cylinder Wasp Major.[3]

# WHEN THE GOING GETS TOUGH . . .

Building aircraft engines is not a business for the faint of heart. The demands for performance and reliability are unlike almost any other required of a manufacturer.

[2] Renstchler's description of Brown's illness and death is the most touching part of his private memoir. He recalled that he had Brown come to his Florida vacation home for several weeks in 1939 where they made plans for the reorganization of the company. When they returned to Hartford, Brown's condition worsened, and Rentschler was under pressure to do something: "I refused to have anything to do with initiating important changes as I knew it would certainly indicate to him (Brown) pertinent things with respect to his own illness."

[3] Old Pratt hands love to play a trick on new folks. "What was the other Pratt engine named after an insect besides the Wasp and the Hornet?" The answer is the Yellow Jacket, the name given an early 1930s liquid-cooled engine. The one example of the Yellow Jacket resides today in the Pratt & Whitney museum next to the H-3730, the engine Hap Arnold and Frederick Rentschler killed in 1940.

The Granville Brothers' Gee-Bee racer.

One story from the late 1930s illustrates what a difficult and frustrating task it can be and how painstaking research and development can find answers in what may look like the smallest of modifications.

Hobbs and his team had Pratt focused on key programs, and the new 1830-C Twin Wasp seemed to be doing well as the Pratt team pushed its performance to meet the requirements of tough, wartime duty they saw in the future. Master rod bearings began to seize during tests simulating high-speed dives, and so something better than a steel-backed, lead-bronze bearing had to be found.

Wright Parkins, a brillant young engineer who would rise to head the company one day, was put in charge of a program that had virtually the whole engineering department working on the maddening, subtle bearing problem.[4]

They tried adding tin to the alloy and pure silver and a layer of lead that was only three one-thousandths of an inch thick backed by silver and copper. It seemed to work, but then more tests showed that for some reason the lead would disappear from the bearing after only a few hours running. It was back to six-day weeks and 12-hour days. Finally, it occurred to the Pratt engineers that maybe it was not mechanical erosion at work. Maybe it was corrosion. They remembered that during engine operation organic acids would form in lubricating oil. Could this be the culprit? It turned out it was. But how do you fix it? In effect they had to figure out a way to "inoculate" the lead in the bearing against corrosion. They tried various combinations of gallium, columbium, and indium along with antimony and tin. Indium turned the trick. The bearings held up. Pratt shared this discovery with other engine manufacturers. As power ratings and diving speeds moved up, this new kind of bearing stood up like a thoroughbred.

Over the history of aviation propulsion, both at Pratt and throughout the industry, there are dozens of stories like this. It can be a tough business.

The decks were now clear, and Pratt & Whitney and its people were about to face the greatest challenge ever given the aviation industry: 50,000 aircraft a year.

[4] Around this time Parkins was showing a visitor through a test house. The visitor observed, "Actually, Mr. Parkins, you people are simply trying to contain and control fire, aren't you?" "Yes," the plainspoken North Dakotan replied, "and that's simply all the devil has to do in hell as I understand it!"

# 50,000
## Airplanes

## chapter 3

**ALMOST EVERYONE IN AVIATION** remembers President Franklin Roosevelt's 1940 call for 50,000 airplanes. What may have been lost to some degree in the postwar euphoria over one of the greatest industrial efforts in history is the state of the industry he asked to do that job.

## STARTING FROM BEHIND

The military aviation budget for the fiscal year beginning July 1, 1938, was $122 million, including airplanes, engines, the whole kit and caboodle. Aviation employed 36,000 people and had industrial assets worth about $125 million. The 1950 Pratt & Whitney company history points out that there were more people working in the hosiery industry than in aviation. Pratt's situation was typical, perhaps even a tad worse. At the close of 1938, Pratt had orders for $8.7 million in engines. That was only enough to keep the 3000-person workforce busy through May 1939. The Army had decided that the Allison V-1710 engine would power its fighters, beating the Pratt R-1830. It looked like military shipments would end in June of that year unless something happened.

President Roosevelt visits Pratt & Whitney in 1940 with his call for 50,000 airplanes a year. Frederick Rentschler is seated directly in front of the president.

What happened were the Munich Crisis and its aftermath. Late in 1938 France, whose industry had languished in the political morass of the Third Republic, started looking for aviation suppliers in the United States. It ordered $2 million worth of Twin Wasps. In 1939 France

Production of engines like the Twin Wasp reached unimaginable levels. Pratt & Whitney and its licensees built 363,619 engines during the war years.

increased its orders to a total of $85 million and agreed to finance a 280,000-square-foot addition to the East Hartford plant. The company hired some 3000 new people to meet the demand. In 1938 Pratt had shipped about 120,000-horsepower worth of engines per month. By December 1939 that figure was 400,000 horsepower a month. After the fall of France in 1940, Great Britain ordered Wasps for training planes and then Double Wasps and took over most of the French business. It also financed a 425,000-square-foot addition, and Pratt began work on another 375,000 square feet, the so-called "American addition" complementing the "French Building" and the "British Building."[1]

Those orders got Pratt & Whitney up to speed as America came closer and closer to war. It was a huge advantage. Also, the work of Rentschler, Horner, Gwinn, and Hobbs and their team in straightening out development programs in the late 1930s was paying off. Looking back on World War I, Rentschler believed that any nation needed a "going article" when war came. The Liberty engine had required a huge effort and had done little if anything to help win the war. The Wasp, the Twin Wasp, and the Double Wasp were all going articles by 1939.

Noted aviation historian Walter Boyne points out in his history of the air war, *Clash of Wings*, that the Germans and Japanese had superior aircraft at the beginning of the war, but were at the top of their development cycles. The U.S. industry was still on an upward climb with newer, better designs that would pay off handsomely beginning in about 1943. It is interesting to note that the Japanese and Germans were still relying on Me109s and Zeros to a large degree at the end of the war, albeit much improved. U.S. pilots fought with Corsairs, Hellcats, Thunderbolts, and Mustangs that were in early development in 1940–41.

Of course, the other key was the huge industrial potential of the United States. In 1940, most people gasped when Roosevelt called for 50,000 airplanes. By the end of the war, American industry could produce 100,000 airplanes a year.

[1] Two other buildings, North War and South War, were added for additional office space. They were called "temporary structures." Those temporary buildings were not demolished until the 1990s. North War in particular posed a problem. Its frame had been built with huge laminated oak beams because steel was in short supply. When the wrecker's ball first hit the old building, it bounced off the oak. Some of the beams had to be blasted apart with small explosive charges—so much for temporary wartime construction.

# THE SHADOW (PLANT) KNOWS

Pratt & Whitney's war story illustrates how this was accomplished. Even with all of those new buildings, there was no way Pratt & Whitney on its own could meet the requirements for global war. Rentschler had always believed in subcontracting as much as 50 percent of the work on an engine to avoid Pratt having to go through the peaks and valleys of order cycles all by itself. Now that would be taken to the next step.

William Knudsen, director of what would become the War Production Board, had convinced a now fully recovered George Mead to take over supervision of the board's aeronautical tasks. Mead met with Rentschler, who was convinced that the best solution to the rapidly increasing demand for engines was to let Pratt & Whitney expand as much as practical and then get the auto industry into the business. But not like in World War I. Rentschler and his team wanted the automobile folks to build proven engines from Pratt, Wright, and Allison, in plants that duplicated the techniques and facilities of the aircraft engine makers, the "shadow plant." Pratt wanted to start with Ford.

Knudsen predicted that Ford would be building aircraft engines within six months. Rentschler said it would be more like 18 months to two years before Ford could hit quantity production. A Ford team came to East Hartford and went through the plant. Later Edsel Ford himself came for a visit. After seeing how aircraft engines were built, Charles Sorsensen, Ford's general manager, said, "What we want to do and propose to do is duplicate your whole operation from the floor up." Ford did a top-flight job, but Rentschler was closer to the mark than Knudsen. Ford hit 400 R-2800s a month in 22 months exactly.

Similar deals were made with Buick to build the R-1830 and R-2000 Twin Wasps. Chevy also made the Twin Wasp and the Double Wasp. Nash-Kelvinator got Double Wasps. Continental Motors and Jacobs Aircraft were tasked with the single-row Wasps and Wasp Juniors, primarily for training aircraft.

The War Department had encouraged Pratt & Whitney to consider building an engine plant somewhere in the Midwest to diversify the aviation industry, which was so concentrated on the two coasts. Also, Pratt & Whitney, the other United Aircraft companies, Colt Firearms, Remington Arms, Winchester, and Electric Boat had soaked up virtually all of the available workforce in Connecticut and Southern New England. The site of an old racetrack near Kansas City, Missouri, was finally chosen, and ground was broken on July 4, 1942, at one of the darkest times of the war years. The plan was that the Kansas City plant would be dedicated

" By the end of the war, American industry could produce 100,000 airplanes a year. "

# HARNESS YOUR PATRIOTISM—BUILD FOR VICTORY

# AIRCRAFT JOURNAL

### Official Publication United Aircraft Club, Inc.

| Volume 2 | APRIL 15, 1942 | No. 4 |

## Lady of the Lathe

Photos by Ivan

**SOLDIERETTE OF PRODUCTION** is the new important role to be filled shortly by Miriam Greenleaf of Vinal Haven, Me., who is being shown the intricacies of lathe adjustments at the Defense Training School of Pratt & Whitney Aircraft by its supervisor of training, William F. Grier. Miss Greenleaf is one of about 50 girls who have begun training as machine operators, enlisting in the factory ranks for the duration.

## HSP Forms Committee For Production Drive

### Organize Plans to Cooperate With Donald Nelson's U. S. Program

A Joint War Production Drive Committee has been formed in the Hamilton Standard Propellers Division of United Aircraft Corporation to operate under the plan sponsored by Donald M. Nelson, Chief of the War Production Board.

The committee is composed of three representatives of management and three of the employees. They are: P. A. Azinger, W. T. Beebe and N. F. Decker for management; J. R. Hooker, B. W. Sierpinski and J. W. Uccello for the employees.

Members of the Suggestions Sub-committee are: P. A. Azinger, A. F. Mannella and F. L. Woodcock for management; and B. W. Sierpinski, William Toth

(Continued on Page Seventeen)

## GLEE CLUB HELPS SELL WAR BONDS

Musically speaking, the United Aircraft Men's Glee Club played a leading role in the successful War Bond rallies held earlier this week in Hartford and New Britain. Several thousand persons heard the rousing songs of the engine builders.

Participation at the Bushnell Memorial and Stanley Arena rallies represented a special honor for the club. The men were the only amateurs appearing at either program; nationally famous 'artists of stage and screen added glamour to the events.

A unique feature of the club's appearance was the warm applause they received when in his introductory remarks, Barry Wood, master of ceremonies at Bushnell Memorial, made reference to the winning of the "E" pennant by Pratt & Whitney en-

## Attention! ... Mr. Hitler

The sweeping response with which the Hamilton Standard Propeller and Pratt & Whitney Aircraft Divisions answered the latest War Bond Drive is compellingly reflected in the percentages of the employees signing for payroll deductions.

Deductions for the entire United Aircraft jumped from $215,000 to $525,000 monthly. Both Hamilton plants increased total monthly deductions from $28,000 to $78,000. At the Aircraft the rise was from $163,000 to $599,000.

Hamilton's Pawcatuck Division led the parade with virtually

(Continued on Page Seventeen)

# Women Begin Training As Machine Operators

## Flier Attacks 6 Jap Planes, Dies a Hero

### By NELSON GOODYEAR, JR.

One of the few Army pilots who was able to get off the ground of embattled Hawaii, while the earth was being rocked with the violence of bomb explosions, Lieutenant Gordon H. Sterling, Jr., formerly of Dept. 30-2, P&WA, lost his life when engaging a flight of six Japanese planes, damaging or destroying one before his own was attacked from the rear and shot down by the enemy.

On March 5 in Boston, General Sherman Miles, commanding officer of the First Corps Area, presented the Distinguished Service Cross to Gordon H. Sterling, Sr., posthumous award for his son's 'heroism and achievement while

**GORDON H. STERLING, JR.,**

participating in an aerial engagement with Japanese planes over the water adjacent to the Isle of Oahu on December 7." The ceremony was broadcast over the Colonial network and carried in the Hartford area by Station WTHT.

Mrs. Sterling, the flier's mother, said, in an interview, that Gordon had been interested in flying since his childhood, and that he had hoped to make a lifelong career of service in the United States Army Air Corps.

After graduation from the William Hall High School in West

(Continued on Page Twenty Two)

## Keep It Flying

### Navy Awards E Pennant To Aircraft

"E" for Excellence, says the United States Navy.

"E" for Engines, echo back 26,000 Pratt & Whitney Aircraft employees.

Expressed briefly this was the spirit of the colorful ceremonies held the afternoon of March 24 on Rentschler Field when Rear Admiral John H. Towers, chief of the Bureau of Aeronautics, presented the Navy's cherished "E" pennant to these same employees in recognition of their vigorous war effort. Virtually all of the employees surrounded the speaker's stand for the historic

(Continued on Page Eight)

## Commuting UAC Employees Wear Out 17 Tires a Day

Employees of the East Hartford divisions of United Aircraft who drive their own cars to work are wearing out tires at the amazing rate of nearly three hundred fifty thousand miles a day. This was one of the startling facts revealed in a transportation survey recently completed by the corporation.

Figuring an average life of 20,000 miles per tire, these figures indicate that 17 new tires are being completely worn out every working day by employees of United Aircraft just in driving to and from work in the East Hartford plants. They are particularly significant in view of the recent announcement that in the State of Connecticut the April allocation of retreaded tires is less than nine per cent of last year's sale of new tires. Figured on another basis, this mileage consumes approximately 20,000 gallons of gasoline per day—enough to fly 20 bombers from California to Pearl Harbor every working day.

The survey also revealed that while approximately one half of

(Continued on Page Nineteen)

## HSP LEASES TEXTILE MILL IN TAFTVILLE

The Hamilton Standard Propellers division of United Aircraft Corporation, to take care of greatly increased schedules, has leased the one-story brick building at Taftville, Conn., formerly occupied by the J. B. Martin Company, a textile manufacturing concern, Sidney A. Stewart, General Manager, announced today.

The building, which has been vacant three years, will undergo renovations for manufacturing and assembling propellers. Operations are expected to begin in July and the plant, it is hoped, will get into full production by fall.

## School Fits First 50 Girls For War Jobs

### By NED BENHAM

A program training women to take their places in war industry operating the vital machine tools with which final victory must be produced is now being successfully carried out at the P & W A Training School, Hartford.

Working the second shift there today are some 50 alert, serious women who, in spite of the fact that they have been in training but a short while, are already showing progress satisfactory to H. C. O'Sullivan, Supervisor of Training, United Aircraft, William F. Grier, Supervisor of Training, Pratt and Whitney Aircraft, and James K. Gilman, in charge of the Inspection Training Dept. at the school.

### Entering New Field

Bench work, small parts assembly and inspection have generally been considered women's province. P&WA is proving that machine tools can be added to this list. Who can estimate the number of men that may now be released from production operations for other important work? According to Mr. O'Sullivan, no special privileges are planned for these new trainees. They are on

(Continued on Page Nineteen)

# Jap Students Had to Fight Or Be Shot, Claims E. A. Becker, Globe-Trotter

### By DANIEL V. CALANO

"We do not want to fight you but we must, or be shot!"

Such was the message that Ernest A. Becker, divinity student, globe-trotter, teacher in the Orient, was implored to carry to the Chinese from Nipponese college students a year ago.

Mr. Becker, a student at Hartford Theological Seminary and presently employed in Dept. 50

China. Already the war lords of Nippon have expended vast quantities of life and material in the 'China Incident' and have little to show for it," he emphatically declares.

"Nor can Japan continue her conquests at the present rate, for her internal economy is near the verge of complete bankruptcy," Mr. Becker predicts.

### Cities Lack Young Men

"Japan's military machine has

around. And there are schoolboys playing—plunging bayonets into dummies with all the strength of their eight years!"

In China Mr. Becker saw hundreds of Japanese in a cremated form. In Nanking ashes of Nipponese warriors filled "little white boxes," standing row upon row awaiting return to ancestral burial grounds in Nippon, when the P&WA employee was in solely to the already in-production R-2800-B Double Wasp. But as plant construction progressed, Pratt was ready to bring out the altogether better R-2800-C. A huge gamble was taken. Kansas City would produce the C model, an engine not fully developed yet and certainly not in production, with a brand new workforce that had never been inside an engine plant before.

In 1943 William Knudsen visited the plant still under construction and said, "I am going to be a surprised man if we ever get an engine out of this plant. If we do, it probably won't run." The Missouri workforce and the Pratt management team gave him that surprise. Within six months of Knudsen's visit, they had C models up and running. During the next 18 months, they produced 7,931 engines with 21,506,167 horsepower.

Perhaps the greatest achievement of the war years at Pratt & Whitney and hundreds of other companies was creating a workforce that could build the engines, planes, ships, and weapons that the global war effort needed. In Pratt's case a workforce of 3000 highly skilled people would grow to 40,000 by 1943 in what a company publication described as "a tide of school teachers, lawyers, farmers, laborers, tobacco growers, salesmen, waiters, actuaries, clerks and statisticians that came into the bewildering, one-roofed city of connecting shops and test houses." And many of the new Pratt production people were women, something that had never happened before. One of the keys to success was to get these new people to master a "single-purpose, single-operation" machine. In prewar Pratt & Whitney, craftsmen could set up and run virtually every machine in the shop. In the case of the Kansas City plant, the job was even more daunting. The nucleus of East Hartford people had to train and organize a workforce that eventually came to 24,000. In Connecticut Pratt literally imported people from all over New England. The Maine down-east accent, often with the spice of French-Canadian, was heard everywhere.

Years later, people still remembered the spirit of those days. Eunice Mills had been one of those women a company publication of the time described (rather patronizingly looking back) as "determined to master a Fay or a Bullard as they had once been to bring the fluff to a pie crust": "We had a one-purpose goal, and we developed a wonderful feeling of togetherness," she recalled years later. "Everyone was here to get the boys back home from the service."

The Grumman Hellcat.

# INNOVATION: IN PRODUCTS, IN PRICES

An innovation that gave Pratt & Whitney its great reputation in World War II was water injection. Many a pilot would credit the power boost of water injection in his Double Wasp for getting him out of a tight spot. In 1942 the German Focke-Wulf 190 was giving Allied pilots fits with its speed, firepower, and superior climbing ability. The Air Corps was looking for an answer, and Pratt & Whitney had one in the works thanks to some experimental work done in 1938, once again showing that early R&D could pay off when most needed. Pratt & Whitney had looked into a system that injected water or an alcohol/water mix into an engine's air supply. The liquid evaporates and cools the air, increasing its density. This allows for more air to go into the combustion cycle and more fuel to be burned, increasing power without excessively high temperatures. With the call for more emergency power from the Air Corps, in just three months Pratt engineers doped out a system of water injection. It could boost the R-2800 by several hundred horsepower in an emergency and kick up top speeds by as much as 40 mph. By 1943 the system was in service. Water injection became standard equipment on Corsairs, Thunderbolts, and Hellcats as those planes and pilots mastered their German and Japanese opponents.

> **I yelled over the sound of the engines that if anyone wanted to bail out, now was the time.**

During the war, pilots often visited Pratt & Whitney, sometimes as part of a war bond drive. They always credited their Pratt & Whitney engines with pulling them through. The story of Captain David Burton and his B-24 crew was typical as reported in the company newspaper of the day. Going in for a bombing run antiaircraft fire got one engine. By the time he dropped his bombs, another had to be shut down. Captain Burton picked up the story.

> We set about making the last two Pratt & Whitney engines take us as far as they could. Over the next few hours the engines responded every time we applied full power. I yelled over the sound of the engines that if anyone wanted to bail out, now was the time. The consensus was, "Hell, no we will stay with the plane and those two beautiful Pratt & Whitney engines." As we finally neared our field, we knew we didn't have the luxury of making more than one approach. All the crew took crash-landing positions. The B-24 and its two Pratt & Whitney engines glided down just as the other planes had done. Our ground chief couldn't believe what we had done to those engines. They were so hot when we shut down some of the parts actually melted. That night as we sat around bragging about our mission, we paused and toasted Pratt & Whitney and those wonderful men and women who make those magnificent engines.

Even though Pratt & Whitney engines, whether made in East Hartford or at Ford in Dearborn, were identical, flight crews are a superstitious lot. Bob Rosatti, a young aerial gunner in a Navy Privateer, recalled crew chiefs on Pacific islands checking crates and always trying to cadge a "real" Pratt & Whitney engine as opposed to one built by a car company. By the way, Bob got a job at Pratt & Whitney after the war and rose to become a senior vice president and president of International Aero Engines.

As the war effort geared up, Rentschler felt comfortable leaving more and more of the day-to-day duties to Horner, Gwinn, Hobbs, and others who had risen to the top echelons of the company. His forte had always been long-range planning, and even in 1940 he was looking ahead. The last thing this dignified, private man wanted was the label "war profiteer" that had hit people he knew well during World War I. And the accusations and innuendos of the airmail hearings still stung. He was also already concerned about the postwar era when the huge aviation industry that was growing before his eyes would face a difficult and trying readjustment. Always an astute businessman, Rentschler realized that limiting profits somewhat now and putting away funds for postwar adjustment made good sense and could avoid postwar scandals. He was also betting that increases in volume would lead to efficiencies and lower costs, even with material and tooling costs going up rapidly.

The contrails of P-47 Thuderbolts defending 8th Air Force bombers.

He proposed that United Aircraft voluntarily limit its profits to what he called "something which we could all honestly agree was completely conscionable and above all reproach." Undersecretary of the Navy James Forrestal bought the UAC proposal in 1941. He used it as the basis for the 1942 Renegotiation Act, which covered the whole defense industry. In 1946 Forrestal wrote to Rentschler: "I doubt if we could have got through the renegotiation plan without the cooperation of you and a few others like you in business who recognized its fundamental soundness and the protection it would afford after the war."

# SUMMING UP

During the war, profits on government work ran between 3 and 1.5 percent on an after-tax basis. But the volume of business was incredible. Pratt and its licensees built 363,619 engines during the war years totaling 603,814,723 horsepower, about half the horsepower required by the American air forces. Wright had supplied about 35 percent with the remaining 15 percent or so split between Packard, Allison, Lycoming, and others. Pratt engines had powered some 70 different military aircraft models. Besides the Corsair, Hellcat, and Thunderbolt, there were the Liberator, the Devastator, the Catalina, the Ventura, the Marauder, the C-46, C-47, and C-54 and even the AT-6 Texan/Harvard that taught thousands of pilots to fly.

Another measure is how much engine power had grown. In 1938 the most powerful Pratt engine, the Twin Wasp, delivered about 1200 horsepower. In 1945 the water-injected Double Wasp could do over 3000 horsepower in combat, and the Wasp Major reached a 3500-horsepower combat rating about six months before the end of the war, although it never saw active service.[2] Rentschler had missed the timing on that one. Five years was not enough to develop the Major for combat, especially when the Twin and Double Wasps were being constantly improved and filling in well. The Wasp Major would earn its wings in the postwar era.

The company had grown from 3000 to 40,000 men and women, the first time women had entered the Pratt manufacturing workforce in any kind of numbers.[3] From 1939 to 1944, they had worked 48 to 50 hours a week minimum with no vacations or shutdowns. Pratt had grown to well over five million square feet of factory and office space. Yes, the government had paid for much of that. But the war was over, and what would they do now with that magnificent workforce and all that capacity?

The B-24 Liberator, the most-produced heavy bomber of the war, that flew for many Allied nations as well as the United States.

---

[2] A complete list of Pratt & Whitney engines and the aircraft they have powered is included at the end of this book.

[3] In language that seems pretty condescending in today's world, a 1950 company history described this as "a distracting army of women where there had never been women before."

# What Do
## We Do Now?

## chapter 4

ON AUGUST 14, 1945, President Truman announced the unconditional surrender of Japan, V-J Day. World War II was over. On the morning of August 15, the government cancelled $414 million in Pratt & Whitney contracts. The order book now stood at the frightening figure of $3 million, about 10 days' worth of work. The workforce that had done such a magnificent job was told to go home. It had already shrunk from its 1943 high of 40,000 to about 26,000. The plants closed for two weeks, and a skeleton crew of some 6000 people took inventory. Upstairs, Frederick Rentschler, Jack Horner, and Bill Gwinn took stock.

## EMPTY HORIZON?

For years after those dark days, Bill Gwinn had a cartoon on his office wall. He and Jack Horner are sitting in a small, becalmed sailboat. Gwinn is depicted studying a chart and asking Horner, "See anything, Jack?" Horner, his hand on the tiller, peers out at an empty horizon and says, "Nope."

Actually, the horizon was quite full. . . of daunting questions. Could Pratt & Whitney get into jets that were the coming thing in aviation? How could it compete with General Electric and Westinghouse that had a head start on gas turbine technology? Did the piston engine have any future at all? If it did, could Pratt compete with its own war production engines that were sure to flood the postwar surplus market? What should be done with the huge production facilities created for the war? What would the overall aviation market be like? Would it fall apart as it did after World War I? Would airlines thrive, but on surplus planes with no new sales? Would personal aircraft finally materialize in those new suburban driveways?

But the biggest question was jets. During the war, the United States had gotten technical information on Sir Frank Whittle's

The mighty Wasp Major.

breakthrough engine in England and started development work. That work went primarily to GE and Westinghouse along with the Allison Division of GM, Allis-Chalmers, de Laval Steam Turbine, and Chrysler. It made sense because GE and Westinghouse and the others knew turbine technology. Virtually no money went to Pratt & Whitney and Wright because their piston engines would win the war and they could not be distracted from that work.

Pratt's only experience with gas turbine engines was a thing called the PT-1. It was not really an engine as much as a test article. It consisted of a two-stroke diesel engine, the exhaust of which drove a turbine wheel connected to a propeller shaft. The idea was to eventually create a turboprop with very low fuel consumption that could work on long-range patrol and bomber aircraft. The effort, begun in 1940, got a little Navy funding, and Pratt worked off and on on the concept, for that was really all it was, during the war years.

> "What would the overall aviation market be like? Would it fall apart as it did after World War I?"

So in the fall of 1945, Pratt's outlook in the jet field was bleak to say the least. The business people gave the company five years to catch up with or beat the competition. After that it might be too late. The engineers said they would need at least those five years to get a "going article" ready. So summer 1950 became the target, but what to do in the meantime?

One break came when the Navy needed additional production of the Westinghouse J30 small, axial-flow turbojet. Pratt got the deal and built 130 of the little engines: its first experience in how to manufacture a jet vs a piston.[1]

# DO YOU HAVE WHAT IT TAKES?

This might be a good place to stop for a moment and explore the issues Pratt faced in learning what was essentially a whole new business. In 1952 Jack Horner lectured at the Industrial College of the Armed Forces on how different jets were from piston engines. First the amount of design work was radically different. The first four years of work in the R-4360 Wasp Major took 730,000 design man-hours. The first four years on what would become the J57 jet took 1,338,000. A jet has an operational advantage because it rotates in one direction and the forces at work inside the engine are usually at a relative steady state, even if temperatures

---

[1] There are basically two types of jet engines, a centrifugal design and an axial design. In a centrifugal system the compressor is a disk with a series of airfoils that pick up incoming air and direct it toward the center of the compressor. Centrifugal force then pushes the air at high speed and pressure into the engine's combustion chamber. In an axial-flow engine the air moves straight through the compressor, with each compressor stage building up more and more pressure.

and pressures are high. This compares with the pounding that a piston engine takes. Horner cautioned, however, that the design of a jet takes so much effort because it involves so much theory and analysis of thermodynamics and aerodynamics. The manufacturing in one sense is easier: "It is unlikely that any part in a gas turbine engine will ever be as complicated to machine and finish as a piston engine's master rod and crankshaft," said Horner.

On the other hand, piston engines were built mostly from castings and forgings, and the new jets required large amounts of sheet metal work. Horner said the jet only became possible because of the development of alloys for that sheet metal that could handle higher temperatures and pressures. And Pratt had to figure out how to weld sheet metal. "With the multiplicity of joints in sheet metal parts of a jet, the distribution of stresses is one of the most important considerations. A weld becomes an actual design factor rather than a mere fastening device," Horner said.

He referred to many of the issues in converting to jets: things like relatively large diameter parts with very thin walls and all of the compressor and turbine components and airfoils with "a great variety of aerodynamic shapes of such awkward dimensions that our designers often complain that they have neither a beginning nor an ending."

He called these "odd horses and peculiar cats in strange contrast to the piston engine's comfortable old forgings and castings which were heavy and sturdy and supplied their own rigidity for machining." And all that sheet metal and oddly shaped stuff needed a lot of tools. To build the little J30, Pratt needed 5250 tools. By 1952 when Horner spoke, the J57 had 20,000 tools.

As 1945 turned into 1946 and 1946 into 1947, however, key decisions had been made. Pratt & Whitney would take five years to get into jets, and, most importantly, it would not catch up with the competition, but would leapfrog them. It was decided that the best way to do this was to come up with an axial-flow engine because axial-flow turbojets looked like they could produce a lot more power than the centrifugal design of the original Whittle. Also for a fighter the axial-flow engine was much slimmer than the Whittle type with much better drag characteristics.

As Horner observed, this would take a lot of gas turbine analysis and theory to pull off. So he and Gwinn and Luke Hobbs went to Rentschler and the UAC board and said they needed a really advanced laboratory facility, something unlike anything that existed virtually anywhere. In 1947 they got the go-ahead for what would become the Andrew Willgoos Gas Turbine Laboratory with test cells that eventually could replicate up to 100,000-feet altitude.[2]

35

---

[2] It would be so named in 1950 after the 1949 death of Andy Willgoos, who died while shoveling snow in his driveway so he could come in to work.

It was one of those "bet-the-company" moves that are so famous in aviation. Horner recalled many years later:

> This laboratory was going to cost $15 million. That doesn't sound like much today but it was an awful lot of money back then. I think the total fixed assets of the corporation was maybe around $10 million and the thought of a $15 million addition to those fixed assets was a pretty tough lump for the directors to go for. I sure admire them for standing up to that one because if it had not worked out they could have been in for very, very severe criticism.

For many years Willgoos was the largest privately owned gas turbine research facility in the world. To get some idea of its scope, the power to run its altitude test cells came from the boiler and engines of a surplus Navy destroyer escort.

# A BUNDLE FROM BRITAIN

Another step forward in learning about gas turbines came in 1946 when the U.S. Navy decided the Rolls-Royce Nene engine was what it needed for the new Grumman F-9 fighter. But the Navy wanted an American version of the engine. The American rights to the Nene were actually owned by Philip Taylor, a former general manager at Wright, but he had no manufacturing capability available. Pratt & Whitney saw an opportunity. It

**" The Comet would knock out American dominance of the airline business. "**

was a deal with the company's oldest customer, the Navy. It involved Grumman, with which Pratt had worked closely for years.[3] It got Pratt more into jets with a well-respected Rolls-Royce centrifugal design years before Pratt's own axial would be ready.[4] So Pratt & Whitney acquired the American rights to the Nene, and Bill Gwinn and a Pratt delegation took off for England to work out details with Rolls-Royce and get specifications and drawings. It would be a daunting task, building a new engine of a type Pratt & Whitney knew little about. And the Navy wanted production engines out of East Hartford by November 1948, 17 months from the time the deal was done with Rolls.

The first Pratt & Whitney jet, with a Rolls-Royce pedigree, the J42.

The engine, dubbed the J42 in the United States, also had to run on gasoline as well as kerosene, requiring a redesign of the fuel pump system. And everything

---

[3] All of the Grumman cats—Wildcat, Hellcat, Bearcat, Tigercat, and Tomcat—were Pratt-powered.

[4] Pratt & Whitney always has had a very healthy respect for Rolls-Royce as a competitor and a collaborator, a relationship that carries on from the days of the Nene license to today's International Aero Engines V2500 program.

in the J42 had to come from American sources, most of which had never built anything for a gas turbine before. All of the British drawings had to be converted into American technical language, and some redesign was required to meet the Navy requirements. But by March 1948 the first Pratt & Whitney Nene/J42 was on the test stand. And by the fall of that year the J42 passed its qualification tests at 5000 pounds of thrust and 5750 with water injection.

And work was underway on a more powerful version of the Nene that became the Rolls-Royce Tay and the Pratt & Whitney J48 Turbo Wasp at 6250 pounds of thrust.

So progress was being made in the gas turbine field with that 1950 deadline looming on what had been Jack Horner and Bill Gwinn's empty 1945 horizon. But what about the piston engines, the Wasp family?

# PISTONS: THEY JUST KEEP ON TICKING AND TICKING AND TICKING . . .

At the end of the war, the British had announced that they were going out of the piston engine business entirely. Jets were the future. The United Kingdom was the only country in Europe, or anywhere else for that matter, remotely capable of competing with the United States in aviation at that time. It would build jets and dominate. The Comet would knock out American dominance of the airline business.

The B-50 Lucky Lady II returns from its 94-hour nonstop round-the-world flight.

The J57 is mounted on the prototype B-52 bomber.

Frederick Rentschler, who had spent the last years of the war and the early post-war period laying out the strategy Pratt and UAC would follow, didn't buy that the piston engine was a dead letter:

> In my opinion we would continue to build and sell piston engines through at least the first five years of postwar for installation in the heavy, long-range bombers and for all transport requirements. I further stated that in my opinion piston engine production for certain types, particularly transports, would continue through most of second five-year period.[5]

---

[5] In later years Horner, who said he considered Rentschler a genius, recalled his management style: "He was very thoughtful and stayed pretty far from operations most of the time and tried to think about where things were going in the future and he was very, very successful in making those judgments. Usually they were not popular judgments so that it was always that he was over here and a 100 people were over there. But he was almost always right and it took not only brains but it took a lot of guts to always be in the minority position. He was an exceptional person and a great leader and a very quiet, very shy person."

As the 1940s moved on, Rentschler was proved right. The mighty Wasp Major, which did not make it into World War II, hit 3500 horsepower in early 1945. It became the powerplant for the Convair B-36 and then the Boeing B-50, the nucleus of the U.S. strategic bombing force in the early days of the Cold War. It was used on the B-35 Flying Wing and on the Douglas C-74 and C-124 Globemasters. And the R-2800 Double Wasp was souped up to push the Corsair and the new Grumman Bearcat above 40,000 feet. The so-called Goodyear Corsair got a Wasp Major, which made it a formidable ship indeed.

Between February 26 and March 2, 1949, a B-50, the Lucky Lady II, made the first nonstop flight around the world using aerial refueling. Its four Wasp Majors had run continuously for 94 hours and 1 minute with no malfunctions.

On the commercial side, it was perhaps this kind of "dependable engine" that vindicated Rentschler's prediction that the piston engine would not fade away immediately. Pratt & Whitney engines were specified for the DC-6, the Martin 2–0-2, the Convair 440, and the Boeing Stratocruiser. Wright did beat Pratt for the Lockheed Constellation and the DC-7 with the R-3350 Turbo-Compound. The engine was very advanced for its time and is sometimes called the ultimate piston engine, but it was highly complex. There are people who say it was the wrong path for Wright (or Curtiss-Wright) to follow. The company did attempt some turbine development but did not make the substantial R&D investments it perhaps should have. It secured licenses for British jets but was out of the engine business by the late 1950s.

But the old Pratt & Whitney philosophy of continually improving production engines until "the last ounce of power and reliability" paid off. Hundreds of DC-4s came out of surplus yards and into airlines after the war. The DC-4 had been designed around the R-2000 Twin Wasp just before the war and did sterling service as a transport. But Pratt had improved the engine significantly by 1945, and in the fall of that year American Airlines made what was then the largest commercial order in Pratt & Whitney history. It wanted 252 spanking new Twin Wasps for its fleet of reconditioned, surplus DC-4s.

It was the C-54 that would also be the mainstay of the Berlin Airlift, the first staredown between East and West in what was being dubbed the Cold War. The airlift not only showed Western resolve in the face of Soviet threats, but what air power could do, peacefully. About 250 planes were flying on average each day, primarily C-54s. Over 321 days they carried about 1.5 million tons of cargo to the cutoff city in 196,000 flights spaced only 90 seconds apart.

That Pratt & Whitney commitment to continual product improvement helped it weather the storm of surplus aircraft and engine sales in the late 1940s. Many airlines bought newer, more powerful, more reliable Pratt engines in preference to the older, war surplus models of the same type. Still, the availability of those war surplus planes cut into business. Smaller operators could do fairly well with an old DC-3 or DC-4. And there were lots and lots of surplus spare parts available. Jack Horner later lamented, "The surplus market took a lot of spares business away from Pratt & Whitney at a time when we really needed a little business to keep going."

# THE JETS ARE COMING! THE JETS ARE COMING!

The big issue was still jets. Would the bet-the-company gamble on a new, powerful axial-flow engine pay off? Could the people of 1950 do what the men of 1925 had done with the Wasp?[6] GE was moving ahead with its designs. Probably the most outstanding was the J47, an axial-flow turbojet that would go on to power the F-86 fighter and the B-47 bomber. Rolls-Royce, Bristol, and deHavilland were working hard on technology that they hoped would give the British industry jet supremacy.

To follow the Pratt & Whitney story, we have to step back now to about 1946. Pratt was learning through the Westinghouse J30 and the Rolls-Royce Nene/J-46 work and was developing its own stable. Little known today, the T34 turboprop was the first axial-flow gas turbine ever designed by Pratt & Whitney. It was aimed at large military air lifters and could produce between 5500 and 7500 shaft horsepower. It first flew in 1947 in the nose of a B-17 test aircraft with its four Wright Cyclones shut down and just the T34 pulling the Flying Fortress through the air. The T34 only saw limited production between 1950 and 1960 and was used on Douglas C-133 Cargomaster, the Boeing Super Guppy, and C-121 military Constellation.

According to Jack Horner's account, Howard Hughes put on a full-court press to get the engine into commercial service on TWA's Connies. Horner and Bob Gross, head of Lockheed, fought the idea, saying the engine was strictly a military powerplant and would not do in airline service. At one point Horner insisted on discussing the idea with a top technical person at TWA to show what a bad idea it was. "No you won't," Horner recalled Hughes saying. "I'm not sure he's going to work for me tomorrow." "Well, so much for Howard," Horner sighed, remembering the eccentric billionaire. The T34 stayed off the Connie.

---

[6] Sadly, by 1950 Frederick Rentschler was the only one of the original troop still active in Pratt & Whitney. Don Brown had died in 1940. Engineering genius George Mead died in 1949. It was only a few weeks later that Andy Willgoos, the designer who "could think with his fingertips," passed away. Charlie Marks, who cautioned workers to "handle parts as if they were eggs, only more carefully," had died in 1943. Gruff old Jack Borrup had retired at the end of the war. It was at this time that Rentschler ruefully conceded that the title "The Old Man" fit him.

The commercial version of the
pioneering J57, the JT3.

41

Although somewhat successful with the T34, Pratt & Whitney did not pursue large turboprops aggressively, and that business would be dominated for years by the Allison (now Rolls-Royce North America) T56. Pratt's turboprop leadership would come from north of the border in a Montreal suburb, a subject that will be looked into later.

Pratt & Whitney was designing another large turboprop at the time, designated the PT4/T57. The Air Force was looking for a turboprop-powered replacement for the B-36. It would be a big, long-range airplane, but relatively slow. However, the revelations in captured German technical documents on the capability of a swept-wing design were becoming clearer. The big slow turboprop bomber went away, and the turbojet-powered, swept-wing B-47 emerged with GE engines. But it was a medium-range bomber, and the Air Force wanted something with more payload and range. It needed a jet-powered, swept-wing aircraft that could approach 600 miles an hour and fly 5000 miles unrefueled at high altitudes. With powerful radar-controlled guns and the emergence of first-generation surface-to-air missiles, speed and altitude were essential. Innovative designs like the Northrop Flying Wing looked promising at first but could not meet the mission. Of course, that concept was the seed that finally germinated into the B-2 in the 1990s.

In 1948 Boeing got a development contract for what would become the B-52, and Pratt began work on an engine for the Stratofortress. The idea was to take the PT4 turboprop core and turn it into a turbojet. Several configurations were tried, and in May 1949 the engineers decided they had to radically redesign the core to get the 10,000 pounds of thrust they wanted along with markedly improved fuel consumption to give the B-52 unrefueled range. The engineering department went on one of those 24-hours-a-day, seven-days-a-week efforts that so many engineering groups in so many aviation companies have had to face over the decades. In exactly 220 days, the reconfigured JT3-A was on the test stand.

The key to the engine's performance was the twin-spool design: a low-pressure and high-pressure compressor driven by matching low-pressure and high-pressure turbines. This boosted power and efficiency tremendously by allowing the engine to develop higher pressures than in a single-spool design. The engine was built around two shafts, one inside the other. One from the low-pressure turbine drove the low-pressure compressor at the front of the engine. The other from a high-pressure turbine drove the high-pressure compressor just aft of the low-pressure unit. Another innovation was the wasp-waist design of the engine. The pinched-in section also helped boost compression efficiency and saved 600 pounds in weight.

The "Old Man," Rentschler, with a J42 mockup.

43

Jack Horner explained the thinking behind the decision to go with something brand new at a time when most jets were producing 4000 to 5000 pounds of thrust:

> It would have been easier to develop an engine of say around 6000 pounds of thrust than it would have been to jump up, as we did, to the 10,000 pound thrust. It would have been an easier job to do a straight, simple single-spool jet than to go with the more complicated two-spool high compressor jet. But with our competition . . . established in jets during the war, we finally decided that we'd better take the jump and go over the top of our competition, both in power and fuel consumption which meant high compression.

He also recalled Pratt & Whitney engineers having to learn how to overcome problems with the new kind of bearings in jets, seal problems, and compressor matching problems, "all the usual things people have with a new jet engine."

Horner added, "They weren't the kind of problems that, in retrospect, were more than you would normally expect. Remember, that's in retrospect. Each time they hit you then, they scared you to death."

The JT3A, which in military parlance became the J57, was indeed a world beater. Eight of them powered the B-52, and the J57 in the F100 Super Sabre made it the first fighter to achieve supersonic speed in level flight in May 1953. By then the Pratt team had won the 1952 Collier Trophy for its J57 work, and the goal set out in the uncertain days at the end of the war had been met.

The engine, in one sense, can be seen as another Wasp. It was a thoroughbred and clean as a hound's tooth and could be adapted for many installations. The J57/JT3 was followed by a scaled-up J75/JT4. Those two engines powered all of the century series fighters of the 1950s with the exception of Lockheed F104. Besides the B-52, the JT3 also powered the KC-135 jet tanker and even the U-2. On the commercial side, the JT3 opened jet travel on the Boeing 707 and the Douglas DC-8. Those airplanes also later used the JT4 for some applications. Author Bill Gunston in his *World Encyclopedia of Aero Engines* described the JT3 as "probably the most important engine in the world since 1945."

In the late 1950s the JT3 was improved once again with the addition of a fan at the front of the engine to markedly increase airflow, what became known as the turbofan. The JT3D's military version was the TF33. Thrust increased 35 percent to an eventual maximum of 21,000 pounds. Fuel burn dropped 15 to 22 percent, and takeoff noise declined 10 decibels. The soundness of the basic design concepts behind the J57/J75 family proves out even today. The B-52s were re-engined with TF33s and still provide sterling service, and there are still 707s and DC-8s all over the world in secondary cargo service.

Pratt & Whitney was becoming the dominant player in the jet engine field, but its competitors were in no way ready to wave a white flag. GE had had success with the J47 on the F-86 and the B-47. It built on that and came roaring back with the

J79, an outstanding axial-flow turbojet with 17 compressor stages. It powered the McDonnell F-4 Phantom, the North American A-5 Vigilante, the Lockheed F-104 Starfighter, and world's first supersonic strategic bomber, the B-58 Hustler. The J79 pioneered the concept of variable compressor stators. The engine's inlet guide vanes that direct air into the engine and the first six sets of stators (nonrotating compressor airfoils) could be adjusted to maintain the correct airflow angles throughout the changing flight envelope. The J79 had a long and successful production run and was arguably the engine that made GE's reputation in the 1950s and 1960s, the cornerstone of its success in the decades to come.

> " Pratt & Whitney was becoming the dominant player in the jet engine field, but its competitors were in no way ready to wave a white flag. "

GE was much less fortunate with its attempt to build off J79 technology and create a commercial engine, the CJ805, a turbofan with the fan placed aft instead of in the front of the engine. GE proposed this engine for the original KC-135 when the increasing weight of the tanker seemed to reach the limits of what the JT3 could lift. Pratt & Whitney responded initially with a water/alcohol injection system for the JT3 and followed on with the JT3D turbofan, blunting the GE effort. Also, the development of the J75/JT4 in this period was partially a response to the success of the J79. The CJ805 would find a home on the Convair 880 and 990 aircraft, but its performance was not good, and the Convair aircraft were poor sellers.

The JT3D was the company's first turbofan and was not only a response to GE but to the Rolls-Royce Conway engine, the first turbofan of any kind to go into airline service. The principle, simply stated, is that a turbojet accelerates a specific amount of air very fast. A turbofan accelerates a much larger amount of air more slowly. A fan stage at the front of the engine generates much more airflow. Some of the air, rather than being accelerated through the core of the engine very rapidly and burned, is "bypassed" outside the core. The more airflow, the more thrust, but because you move a good deal of the air more slowly the engine burns less fuel and makes less noise. In the jet age bypass ratios constantly increased. By the turn of the 21$^{st}$ century, engines were bypassing six or seven times as much air as they were using in the core with even higher ratios on the horizon because of increasing environmental concerns.

Recall that the British were devoting all of their energies to jet technology during the late 1940s and early 1950s. The most famous example, of course, was the deHavilland Comet, which had four deHavilland Ghost engines. The little understood problems of metal fatigue in a high-speed, high-altitude pressurized plane led to three crashes of the early model Comet, and the aircraft was doomed as a commercial venture. It would rise again in the late 1950s as the Comet 4, a much different aircraft powered by Rolls-Royce Avon engines, but it could never catch up with Boeing and Douglas. The Avon was also the initial engine for the French Sud-Aviation Caravelle medium-range airliner. Rolls-Royce did have some success with its Conway engine on extended-range DC-8s and the Vickers-Armstrong VC10, although only 54 VC10s were built.

It should be noted that the Bristol Engine Company was doing exciting work in this period. It would become part of Rolls-Royce after the many consolidations of the British engine industry. About the time Pratt was developing the twin-spool J57/JT3, Bristol was working on the twin-spool Olympus engine. It first ran in 1948 and went into production in 1954, only ever so slightly behind the J57/JT3. It was constantly improved, and the Olympus, of course, became the engine for the supersonic Concorde. It put out 38,000 pounds of thrust from a design that started out producing 11,000.

# MOVING ON, SADLY

By the mid-1950s all of those plans that Frederick Rentschler had laid out at the end of World War II seemed to have fallen in place. In five years Pratt & Whitney had developed a world-beater of a jet engine. Its piston engines had lasted in the marketplace, and the transition from them to JT3s looked ensured. Almost all of the top fighters were Pratt-powered, and the company had grown back from V-J Day to employ some 35,000 people. But Frederick Rentschler, who used to love to play a fast set of tennis with people like Don Budge, Bill Tilden, or Val Yavorsky, was in failing health. His associates noted that he seemed to tire more easily after the death of his wife, Faye, in 1953. He began to lose weight and to age. In February 1955 he was told he had incurable cancer. He spent the last months of his life at his home in Boca Raton, Florida, with his family and died on April 25, 1956.[7]

Frederick Rentschler, once dubbed "Mr. Horsepower" in a *Time Magazine* cover story, was gone, as were the rest of the founders of Pratt & Whitney. They had built a great company on simple principles that would be tested again and again in the years to come.

---

[7] Rentschler's West Hartford, Connecticut, estate is today the home of the Renbrook School. A Wasp engine stands in its lobby.

# Forging Ahead...
## Right into Space

## chapter 5

**DESPITE ITS SUCCESS** with the J57/ JT3 and J75/JT4, Pratt faced a dilemma as the 1950s drew to a close. What was its future? The issue could be called the space race, the missile gap, the Sputnik Scare, or the coming of push-button warfare. Pratt & Whitney built engines for manned aircraft. Were there going to be any more of them or at least enough to keep the company going? Looking back, the concern might seem a bit ludicrous, but it was very real in the late 1950s.

Wright Parkins, who had led so many of the company's key developments, summed up the concern in a speech to Pratt executives in June 1959. He said that the company would have to move from one dominated by engineers and production to one that had more and more scientists and research. Parkins conceded that perhaps United Aircraft, as some industry analysts observed, was "too much production and not enough scientific minded for its own good."

He summed up the fears by observing that Pratt had a 30-year run in piston engines. It had been in the turbine business only 13 years, and "unless the military trend changes, its end is in sight."

An engineer to the marrow of his bones, Parkins seemed very uncomfortable with the idea that aviation was turning into aerospace led by "a small exclusive fraternity of technical intelligentsia." But he and the rest of the Pratt leadership were not ready to throw in the towel:

The site of the new Florida Research and Development Center.

Any outfit that could make a complicated device like a turbine behave as only we have done, could also do the things with rocket engines that are only being talked about today....the Lord has blessed us with engineers who, within the realm of reason, can make anything scientists dream up work.

So, not unlike the period after World War II, Pratt & Whitney sought to reinvent itself yet again. The journey would take it far from its Connecticut roots geographically and technically. Some roads did not lead to great success. Others did.

First of all, there was the issue of those scientists that Wright Parkins was not particularly fond of. To be in the space business, a company needed them, and Pratt & Whitney had none to speak of. That would be fixed. In fact starting in about 1958 and through the next five years, United Aircraft and Pratt & Whitney would spend about $160 million to get into space.

# WHAT IS A CHEMICAL SYSTEM?

In October 1958 a new company was formed in Menlo Park, California, called United Research Corporation. It had one employee, retired Air Force General Donald Putt. He had one task: get Pratt & Whitney into solid rocket motors. Over the years, the organization would have a few different names: United Technology Company and finally Chemical Systems Division.

Back in 1958 Chemical Systems was going up against Aerojet, Thiokol, Rocketdyne, and Hercules in rocket propulsion, all well-established outfits. Don Putt would have to move fast. One of his first steps was to hire key people, some heavyweight scientist/engineers. Barney Adelman, Herb Lawrence, and Dave Altman had been working on the idea of segmented solid rockets but needed money to continue. With General Putt's reputation and Pratt's financial backing, they signed on and brought more experts with them, giving the new organization credibility.

It was supposed to be primarily a research house, but the newly formed team believed strongly that it had to go beyond that, especially in the field of propellants.

The first buildings go up at FRDC.

In a 1983 interview for the company archive, Altman explained, "The controlling technical element of the solid rocket is the propellant, not the mechanical engineering of the hardware.... You need a single, coordinated, and completely independent capability."

So by late 1959 the one-man organization had grown to about 150 people and soon would have a 3100-acre site near San Jose known as Coyote Canyon to develop and test solid rocket motors.

Solid rocket motors begin in essence as a rubbery chunk of both fuel and oxidizer. They do not need an outside source of oxygen for combustion. After the stuff is mixed into that rubbery hunk, it is cast into the rocket case and glued to the walls. The propellant has to be able to withstand tremendous stress when it is burned and must burn in a very precise way. Also, the customer has to be able to store a solid motor for as many as 10 years in a missile application.

> **Some roads did not lead to great success. Others did.**

Another issue is size. A large rocket needs lots of propellant and can weigh as much as 500,000 pounds, 90 percent of that propellant. A vehicle that size with its propellant all in one piece can be mighty hard to transport and handle. The Chemical Systems team came up with the idea of segmentation, building individual power units that could then be assembled on a launch pad. And you could vary the number of segments depending on power requirements.

The frantic work and investment that had started in 1958 paid off in 1962 with a contract to build solid rocket boosters for the Titan III family. The first flight of a Titan III-C was on June 18, 1965, beginning a series of 65 flawless launches that went on until the end of the Titan program in 1982. The propellant was synthetic rubber with aluminum additives for fuel and ammonium perchlorate as the oxidizer.

Looking back, it could be argued that the Titan success was a mixed blessing. In the early years of Chemical Systems, the Titan was so important that the company was discouraged from seeking other work. When the Titan program ended, Chemical Systems was a "one product" organization that had lost its product.

The company did move on, becoming involved in the Scout, the Inertial Upper Stage, and the Trident, Minuteman, and Tomahawk programs. It also was involved in refurbishing the solid rocket boosters for the space shuttle through an organization known as United Space Boosters (USBI).[1]

---

[1] USBI brought a new dimension with it, the UTC Freedom and UTC Liberty, the solid rocket retrieval ships. Around the company they were often referred to as "Harry Gray's Navy" in reference to United Technologies Chairman Harry Gray, who often featured the vessels in corporate presentations.

Although innovative and technically proficient, Chemical Systems never achieved the vision of its early team. Analysts later opined that because it never had the capability to be a full-system integrator it could not grow as much as had been hoped. With the end of the Cold War and a dwindling number of programs to go after early in this century, as well as antiquated facilities that were posing environmental, safety, and customer concerns, the company phased out and closed Chemical Systems over a period of years.

But rockets were still part of the Pratt & Whitney story.

# GATORS, RATTLERS, AND A BLACKBIRD

As the 1950s progressed, Pratt & Whitney was facing problems with its home base in Connecticut. The plants and offices were crowded, and a lot of space had to be leased. Outdoor testing in the area affectionately (or not!) known as the Klondike in East Hartford was drawing noise complaints. It was getting harder and harder to find qualified personnel in the region with not only the demand from Pratt & Whitey but all of the other defense companies located in Connecticut and southern New England.

The biggest concern, however, was security. Pratt was working on some supersecret programs that would be very hard to keep supersecret when neighbors' backyards came right up to the fences of the East Hartford campus. One was the J58, a turbojet originally intended for a Navy interceptor with Mach 3 dash capability. Another was a radically new engine fueled with hydrogen designated the 304. It was intended for a high-flying, incredibly fast reconnaissance plane being developed by Kelly Johnson and his team at the Lockheed Skunk Works. That program was called "Suntan." There was another, dubbed "Shamrock," that involved converting a J57 to burn liquid hydrogen. Not only was security an issue with the programs, but handling hydrogen in a crowded metropolitan area was causing some gray hairs to show up on the heads of program managers.

Veteran engineer Dick Mulready recalled his assignment to get the company up to speed on hydrogen as a fuel. He was given the only two books Pratt & Whitney had on hydrogen. On the cover of one was a

A neighbor pays a call.

picture of the Hindenburg airship exploding, and the other detailed a fatal accident in a liquid-hydrogen lab.

In early 1956, the company started looking for someplace that would solve all of these problems. Florida looked interesting: good climate for outdoor testing, lots of space available, and not too much pressure on the workforce. The clincher came when Pratt ran a blind recruiting ad in an engineering magazine. For every respondent interested in aerospace work in New England, 30 replied that they would love to work in Florida. And the state of Florida was enthusiastic about having a high-technology aerospace company move in. In fairly short order Pratt picked out a 7000-acre site in Palm Beach County. It was part of a wildlife preserve on the edge of the Everglades, and so Pratt actually purchased 9000 acres nearby and gave that to the state in exchange for the 7000-acre site it wanted. So the Florida Research and Development Center was set to go. All Pratt had to do was build it and staff it...in the middle of the Everglades.

> **All Pratt had to do was build it and staff it . . . in the middle of the Everglades.**

A handful of East Hartford folks under Chuck Roelke was given the monumental task, and they really were pioneers. Wright Parkins had strictly forbidden any mass transfer of qualified East Hartford folks to Florida. He needed everyone he had in Connecticut. And then there was the little matter of that swamp. Ned Benham, the first public relations manager for the place, used his literary powers to describe the site as "a paisley pattern of pines, palmettos, palms, ponds, and game paths, populated by deer, wild boar, alligators, turkey and quail."

Roelke and his small team were undaunted, or at least that is what they said to each other. In fact, flying into West Palm Beach in the fall of 1956 they were greeted by a rainbow over the airport. They all agreed it was a good omen, and the rainbow became the early symbol of FRDC.

The Bee Line Highway that runs by the site today did not exist, and the building of it was a task in itself. The highway passed through the menacingly named Loxahatchee Slough that needed tons and tons of fill to cross, and according to legend swallowed one or two bulldozers during construction.

The actual site of the plant and test stands needed a huge amount of drainage and then fill because the water table was only two or three feet below the surface. And then there were the critters. The fill had been excavated from "borrow pits" on the property that promptly filled with water and became home to alligators, among other fauna. The sand and gravel road that led five miles out to the test stands often became impassable in the early days, forcing engineers to spend the night in the half-completed support buildings that were covered in tarps. One story has it that a test crew arrived one morning to find that a rattlesnake had made it under the tarp and spent the night comfortably curled up in a wastebasket. Company lore is filled with stories from later years of people finding alligators sunning themselves in the by-then paved parking lots. And the cafeteria patio had a "do not feed alligators" sign. Somehow it all got done, and on May 27, 1959, the FDRC was formally dedicated.

Bill Gwinn, president of United Aircraft and a Pratt Rat since the age of 19, sounded the company's credo at the dedication:

> In aviation, success today results from the decisions you made five, ten or even fifteen years ago. We first began picturing these facilities four years ago. We knew that they would be costly and would require many millions of dollars of our funds. But we are in the aeronautical power plant business for keeps. Our experience over many years has taught us that no matter how rosy things look at the moment, you had better be planning two or three power plant generations ahead if you want to survive in the field.

Even before the formal dedication, work had been going on. The hush-hush 304 hydrogen engine had first run in the fall of 1957 as part of Suntan.[2] Air Force interest in the aircraft, designated the CL-400 by Lockheed, had started to fade. It did not have sufficient range, and the hydrogen fuel posed safety issues. Suntan faded away completely by the end of 1958. But another part of the Air Force was very interested in the technology. These were the people trying to figure out how to boost satellites into space. Liquid hydrogen was the perfect fuel for that purpose. So the 304 program overlapped with a new one that would become the RL10. It was intended as the powerplant for the Convair Centaur, the second-stage vehicle for the Atlas booster.

66 **These were the people trying to figure out how to boost satellites into space.** 99

Even with the previous work on the 304, development of the RL10 was a challenge. Like so many rocket programs of the late 1950s and early 1960s, it had its spectacular, very bad moments, including an explosion that wrecked a test stand. By the end of 1961, the RL10 team had figured out the problem. It was in the ignition system. By January 1962 the engine passed its preliminary flight test. The RL10 first flew in November 1963 boosting a Centaur. In early 1964 six RL10s of a Saturn S-IV stage pushed a 37,000 pound payload into orbit. Since that time, the RL10 has flown hundreds of successful missions and is still in use today.

Space work in Florida continued with development focusing on high-pressure liquid-fueled engines. From this work came the experimental RL20 and then Pratt's proposed space shuttle main engine (SSME). Having worked on this technology for 10 years, the Pratt team felt confident and was devastated when on

---

[2]The hydrogen came from an Air Products company plant on the site that was publicly called a "fertilizer plant." Old timers recalled that the subterfuge worked well until the first tank car arrived with HYDROGEN emblazoned in its side.

June 28, 1971, Rocketdyne won the program. This basically left Pratt with only the RL10 as a viable liquid fuel engine, although in years to come that would all change. Bitter as some Florida people were by the decision, they did not give up, and over time they were vindicated to a degree. NASA eventually replaced the turbopumps, the heart of the engine, on SSMEs with the Pratt model, which is what flies today.

The 304 "Suntan" engine.

53

# HELLO BLACKBIRD

The other main project at FDRC in the early days was the J58, originally intended for a Navy Mach 3-capable interceptor. But by 1958 the Navy was losing interest in the engine and abandoned the interceptor program. There was no other application, and it looked like the J58 might become a footnote in Pratt & Whitney history. But enter Kelly Johnson and his Skunk Works again. Even though the proposed Lockheed hydrogen-fueled reconnaissance plane had died, the CIA and the Air Force were very interested in a new plane to replace the increasingly vulnerable Pratt-powered U-2. The concerns were, of course, proven out when Gary Powers's U-2 was shot down by a new Soviet SAM. In early 1960 the CIA ordered what was then known as the A-12, which became the legendary SR-71 Blackbird.[3]

But the J58 was in trouble. The engine was unlike anything Pratt or anyone else had ever built, but besides the engineering challenge there was the issue of personnel. It turned out the original decision to staff Florida with almost exclusively new hires was a bad one. The team, skilled as it was, had not had the experience of managing a complex program. In Connecticut a lot of things got done because people had just figured out a way to run things by a kind of osmosis, "the Pratt & Whitney way of doing things." Most of that was not formalized, and the Florida people had not had the opportunity yet to build their own knowledge base. So in late 1962 and 1963 a significant number of East Hartford people headed south. It was a difficult period for both groups, but eventually things worked out, and programs got back on track.

---

[3] A presidential faux pas gave it that name. The plane was supposed to be called the RS-71, but when President Lyndon Johnson revealed its existence at a press conference he inadvertently called it the SR-71. No one seemed inclined to correct him, and the SR-71 it became.

The J58 posed a unique set of problems and even today might be the most technically challenging engine Pratt has ever built. At Mach 3, compressor inlet temperatures were 800°F, turbine inlet temperatures were 2000°F, fuel hit 700°F, and oil 1000°F. Special fuels and oils that could withstand the temperatures had to be developed. But the big issue was materials.

Program manager Bill Brown later recalled, "I do not know of a single part, down to the last cotter key, that could be made from material used on previous engines." The chemistry of the alloy used in the turbine disks had to be changed to give heat resistance and strength. New ways to forge and heat treat the disks had to be developed. Engineers came up with a new alloy to make turbine blades and perfected a way to roll large superalloy sheets into the engine's cases.

Parenthetically, this work and more like it for the RL10 and other projects gave Pratt a huge advantage in materials science. The Florida materials group and those working in Connecticut became world leaders in superalloys, powdered metal, coatings, and very advanced forging techniques. That knowledge base has grown and continues to grow today helping make Pratt & Whitney a leader in materials sciences.

Another major issue was the engine cycle itself. The J58 had started off as a pure turbojet, but for the Blackbird this design would not do. There was not enough power and too much fuel burn. The aircraft had to be able to sustain Mach 3.5 at up to 100,000 feet on long-duration flights. In those conditions there were serious airflow problems. The rear stages of the compressor could not handle the volume of air. This made the first compressor stages and the afterburner inefficient. So the original J58 design was scrapped and replaced with a kind of dual-cycle approach. Up to about Mach 3, the engine operated as a turbojet, all inlet air flowing down the core of the engine. At Mach 3 and above a "bleed bypass cycle" kicked in. A portion of the compressor air is ducted around the rear stages and then reintroduced in the turbine discharge for afterburning. At very high speeds and altitude this gives the engine a 22 percent boost in thrust with no significant increase in fuel burn.

In late 1962 the J58 started flying on the A-12 prototype. The Air Force had also ordered an interceptor prototype, the YF-12A that never went into production. The final version, the SR-71, went into service in 1966. It set just about every record for speed and altitude around, including its last flight. The Blackbird flew from California to Washington, D.C., in 68 minutes and 17 seconds and earned a final place of honor at the Smithsonian National Air and Space Museum.

An RL10, the work horse upper-stage engine.

54

# A SUPERSONIC DISAPPOINTMENT

The next big project for Florida was the engine for the proposed supersonic transport (SST). Probably one of the most controversial and expensive programs of its day, the SST captured everyone's imagination. Those in aviation saw it as the next big step. The budding environmental movement hated it. The bean counters probably fainted when they saw the bills. And there was more than a bit of national pride involved. The French and British were going to build the Concorde, and, by God, the United States was not going to be left in the supersonic vapor trail.

In 1964 Pratt and GE were awarded FAA contracts to start development of engines for an SST while Boeing and Lockheed got the airframe contracts. The FRDC team was most qualified to take on the task and did so with relish. Soon signs and buttons proclaiming "54K A Must or Bust" sprouted everywhere, referring to the 54,000-pound thrust requirement. By March 1966 the first engine, the JTF17, was completed and running, based on technology from the J58 and the TF30 fighter engine. The final test results were compiled by December and submitted with a great deal of optimism. Everyone went home for a well-deserved Christmas break. When people came back to work, they were bitterly disappointed. GE had won. Again those inside and out-side the company were questioning Pratt's future. The SST loss was coupled with the Air Force decision to go with GE for its new C-5 airlifter. GE would get money for supersonic work and for high-bypass engines. They could dominate commercial aviation in the years to come. Poor Pratt & Whitney! As it turned out, the SST never happened, and the C-5 never saw the production that had been expected. And as good as the GE engine for the Galaxy was, there was another plane and engine combination coming: the 747 and the JT9D.

The J58, no other engine quite like it.

55

# ON THE BANKS OF THE BEAUTIFUL CONNECTICUT

Today all Pratt & Whitney engines made in the United States are assembled and tested at the company's Middletown, Connecticut, facility, which sits on a 1000-acre site on the banks of the Connecticut River. But back in the 1950s, the plant was the site of one of the most unique projects in Pratt & Whitney history. It was called CANEL, the Connecticut Automated Nuclear Engine Laboratory.

Its roots go back as far as 1948 when the Congressional Aviation Policy Board said the "nuclear energy propulsion for aircraft (NEPA) should be accorded the highest priority in atomic energy research." This was a time when atomic power was seen as the wave of the future. President Eisenhower had started the Atoms for Peace program. Nuclear-generated power would be so cheap you wouldn't even meter usage.

The idea behind the nuclear engine concept was a bit more sinister in the days of the Cold War. A bomber could stay on station for days on end with no need to refuel, able to deliver instant retaliation. By 1951 a small group of Pratt engineers was working on the concept in a former department store suburban branch. In 1955 ground was broken for CANEL under the auspices of the Air Force and the Atomic Energy Commission. It was a government facility, and Pratt was contracted to run it. Eventually about 2500 Pratt & Whitney people worked there.

GE was charged with developing a direct-cycle engine, that is, the air from the compressor went right through the reactor, was heated, and passed into the turbine. The reactor, in effect, became the engine's combustion chamber.

The concept Pratt worked on was called the indirect cycle. In that, the air was never brought in contact with the reactor. The reactor heated liquid metal, which was passed through a kind of radiator that imparted the heat to the air.

Pratt's engine was designated as the J91. Mock-up hardware was fabricated, and lots of experiments were done with various materials, but no complete engine/reactor was ever built, nor was there ever an airplane built. There were lots of issues about materials, and certainly nuclear safety was a concern. But probably the real reason the nuclear engine programs were cancelled in 1961 was the advent of the ICBM, both land and submarine-based, and the effectiveness of the B-52. This would be the nuclear triad that defined strategic policy through the end of the Cold War.

Pratt & Whitney eventually bought the CANEL site, and over time all of the original buildings were demolished and replaced by new manufacturing, assembly and test, and office facilities.

# GROUNDED ENGINES AND ALCHEMY

During the late 1950s and early 1960s, a confluence of circumstances got Pratt & Whitney into a whole new business: generating electricity. First, the company was looking at ways to leverage its technology and lessen somewhat its dependence

Early fuel cells for the U.S. space program.

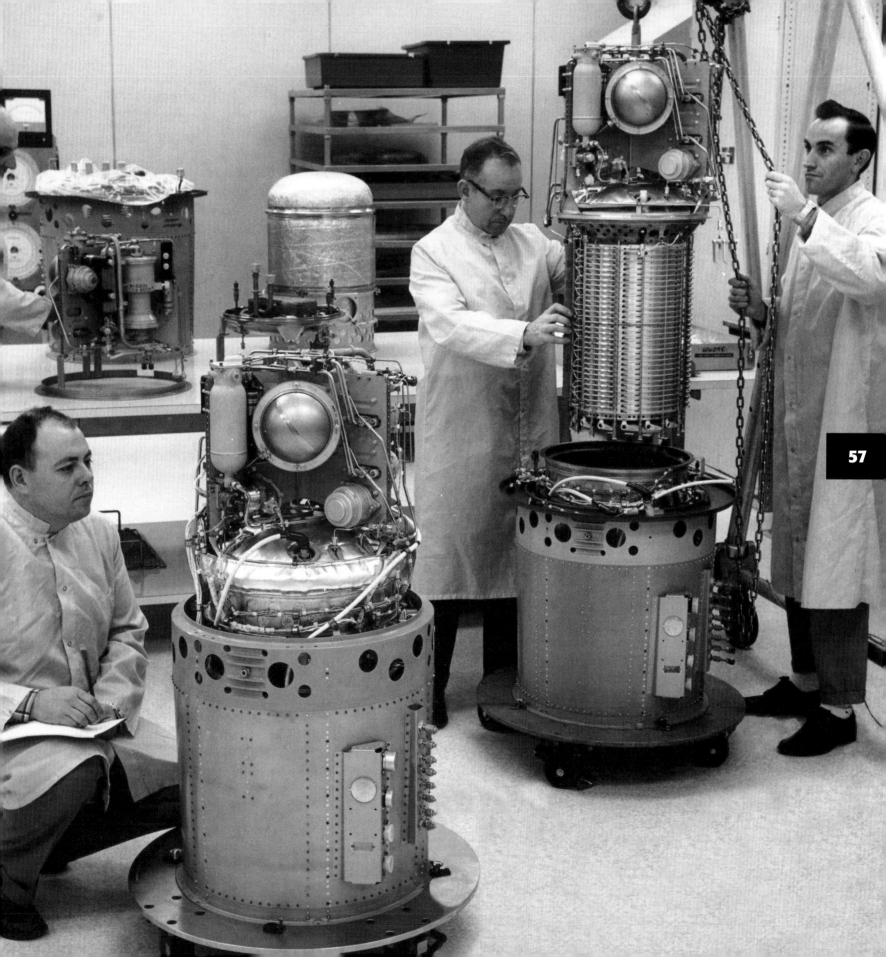

on the military and the notoriously cyclical airline industry. And, as stated earlier, Pratt wanted to get into the space business in a big way. Enter Turbo Power & Marine and Fuel Cell Operations.

Fuel cells might have been the product farthest from Pratt & Whitney's traditional engine business as you could find. They are electromechanical devices that combine hydrogen and oxygen through a catalyst to create electricity, heat, and water. There is no combustion. They are pollution-free and extremely efficient. NASA was looking for a power source for the Apollo moon vehicles. Mercury and Gemini spacecraft had run on batteries. That would not do to get men to the

An FT4 industrial engine.

58

moon and back in a complex vehicle. Since Wright Parkins's recognition of the need for scientists as well as engineers for the company's future, Pratt had been looking at all kinds of new technologies to diversify. Fuel cells emerged from all of that. They were perfect for long space flights, they were clean, quiet, and efficient, and they even produced pure drinking water.

The operation won the contract to provide fuel-cell power for Apollo and then the Space Shuttle Orbiter. Fuel Cell Operations became almost a separate, niche business from Pratt & Whitney and later was even put under the management of Hamilton Standard's space department. NASA was the big customer, but the people at Fuel Cells kept experimenting with ways to translate the advantages of fuel-cell power down to Earth. Under the leadership of the legendary Bill Podolny, they kept getting research grants and small contracts to develop products. The technology, however, took a long time to catch on. Fuel cells are expensive to build and can be temperamental. To bend a phrase, it does not take a rocket scientist to run a fuel cell, but it helps.

By the late 1990s, however, the challenge of global warming and the increasing demand for electricity breathed new life into the technology. United Technologies made fuel cells the cornerstone of a whole new division, UTC Power. That group developed fuel cells for on-site power for building complexes, transportation use, and, of course, space and defense applications. UTC Power has worked with several auto manufacturers on adapting fuel cells for cars. Several hundred buses around the world were running on fuel cells in demonstration projects. Another concept anticipated using fuel cells to provide electrical power to military and commercial aircraft.

# THE ENGINES THAT DON'T FLY

A business more closely related to the Pratt & Whitney core also began developing in the late 1950s and early 1960s—industrial gas turbines. With a growing demand for electric power, gas turbines were a perfect answer for peak winter/summer power demand within the electric utility market. Aeroderivative gas turbines were relatively inexpensive and easy to install compared with large central generation stations. They were very responsive to the changing demand on utility systems and could be supplied at a fraction of the lead time for conventional powerplants. Pratt's JT3/JT4 family of engines was just the right size for the growing electric utility demand. In addition to being an ideal product for the utility industry, aeroderivative engines could be easily adapted for marine propulsion and pumping applications in the oil and gas industry.

The FT8 family of industrial gas turbines can provide almost instant power.

Thus was born Pratt & Whitney's industrial business, Turbo Power & Marine (TP&M), and the FT3 and FT4 families of engines taken from the JT3 and JT4 aircraft engines. Some 1100 of these machines would be sold through the mid-1970s, primarily for power generation. Units also went to cargo ships, U.S. Coast Guard cutters, and some Royal Canadian Navy vessels. Engines also provided power for oil and gas pipelines. One of the biggest jobs came in the 1970s when FT4s were chosen to power ARAMCO pipelines in Saudi Arabia. The FT4 has been one of the most successful industrial gas turbines and gave TP&M a leading market position. By the mid-1970s, however, utility growth rates slowed, and Pratt's resources were stretched to meet new aircraft engine program demands. TP&M's position started to slip when it did not come up with a new product to succeed the FT3/FT4 series. By 1981 TP&M was essentially out of the industrial OEM business. GE and Rolls-Royce on the other hand pursued the business aggressively. This was partially because of Pratt's dominance in commercial engines. The two competing companies needed to find more applications for their gas turbine business, and the industrial segment would do nicely.

Since the late 1970s, GE has been the dominant competitor in the industrial segment, providing both aeroderivative and heavy-duty frame-type gas turbines for the power, marine, and transportation markets.

Not exactly like a phoenix, but with a lot of hard work TP&M, now known as Pratt & Whitney Power Systems, reemerged in the 1990s. Pratt took its most successful engine ever, the JT8D, and adapted it to the industrial market as the FT8. The development effort was painful, and it took some time for the engine to catch on. At times demand for aeroderivative engines seemed to be fading, and Pratt had unsold engines in its warehouses. But by early in this century, the FT8 family was gaining customers. This was partially caused, as in the early days, by an ever-increasing demand for electric capacity to meet peak power demands. The FT8 package was designed in modules that make for efficient cost-effective installation. The Power Systems team also developed a truly mobile FT8 package that can be transported over the road and be up and running within eight hours of arriving at a suitable site. Additionally, Pratt & Whitney Power Systems has expanded beyond the traditional OEM business and is making replacement parts of GE's heavy-duty frame industrial turbines. Providing OEM solutions as well as aftermarket services for non-Pratt & Whitney gas turbines helped make Power Systems one of the fastest growing businesses within the Pratt & Whitney portfolio.

Pratt & Whitney Canada never completely exited the industrial market, and Pratt & Whitney Power Systems ST6, ST18, and ST40 engines, based on Pratt & Whitney Canada aircraft engines, gained market acceptance for applications where less power is needed than the 60 megawatts that the largest FT8 units can provide. Railroad companies experimented with the ST40 engine for locomotives, and marine applications in smaller vessels grew.

So fuel cells and industrial gas turbines, which at one time might have faded from the company portfolio, once again became key items in the future of Pratt & Whitney and United Technologies. Patience is, indeed, a virtue…along with some luck and sweat.

# TIMELINE

**PRATT & WHITNEY EVENTS** & **WORLD AVIATION EVENTS**

## 1920s

**1925** August – Frederick Rentschler and his small team set up Pratt & Whitney Aircraft in Hartford, Connecticut in part of the Pratt & Whitney Machine Tool complex on Capitol Avenue.

**1925** December – First run of the R-1340 Wasp.

**1926** May – First flight of the Wasp.

**1926** June – First run of the R-1690 Hornet.

**1927** May – First flight of the Hornet.

**1928** January – Wasp- and Hornet-powered aircraft perform flawlessly on the first major deployments of the USS Saratoga and the USS Lexington.

**1929** February – Pratt & Whitney Canada begins operation with ten employees.

**1929** Rentschler, William Boeing and Chance Vought lead formation of United Aircraft and Transport Corporation.

**1929** November – First run of the R-985 Wasp Junior.

**1924** Army Douglas World Cruisers complete first round-the-world flight.

**1926** March – Robert Goddard launches the first liquid-fuel rocket on a 2.5 second flight.

**1926** Alan Loughead starts his second company and uses the phonetic spelling Lockheed to end confusion.

**1927** April – First civilian pilots license issued to William P. McCracken, head of the Commerce Department Aeronautics Branch. Orville Wright had turned down the honor because he no longer was flying.

**1927** May – Lindbergh completes first solo transatlantic flight.

**1927** May – First flight of the Being 40A.

**1927** October – Pan Am begins service between Key West and Havana.

**1928** Glenn Martin forms Glenn L. Martin Aviation in Baltimore.

**1928** June – Amelia Earhart becomes first woman to fly the Atlantic.

**1928** December – North American Aviation is organized as a holding company.

**1929** September – Jimmy Doolittle demonstrates instrument flying.

## 1930s

**1930** Pratt & Whitney completes move begun in 1929 to its new factory in East Hartford, Connecticut.

**1930** Wasp Junior-powered Laird wins Thompson Trophy.

**1931** Gee Bee Model Z with a Wasp Junior wins the Thompson. April – First run of the R-1830 Twin Wasp. June – First Twin Wasp Flight.

**1932** July – First run of the R-2060 Yellow Jacket, Pratt & Whitney's first liquid cooled design. The engine never flew.

**1934** "Air Mail Scandal" leads to the breakup of United Aircraft and Transport. Pratt becomes part of the new United Aircraft Corporation that included Hamilton Standard and Chance Vought – Sikorsky.

**1937** September – First run of the R-2800 Double Wasp.

**1939** July – First flight of the Double Wasp.

**1930** May – First flight of the Boeing Monomail.

**1931** October – First Sikorsky Pan Am Clipper enters service.

**1932** May – Amelia Earhart is the first woman to make a solo Atlantic flight. She had become the first woman transatlantic passenger in 1928.

**1933** Boeng 247 first flight.

**1933** July – DC-1 first flight.

**1935** XB-17 first flight.

**1935** December – First flight DC-3 prototype, the Douglas Sleeper.

**1937** April – Sir Frank Whittle successfully runs his first jet engine.

**1938** June – Boeing Clipper and DC-4 make first flights.

**1938** December – Boeing Stratoliner is the first pressurized passenger aircraft.

**1939** James S. McDonnell organizes McDonnell Aircraft Corporation.

**1939** August – He178 first flight with the jet engine developed by Hans van Ohain at Heinkel.

1940 Orders from France and England along with President Roosevelt's call for 50,000 airplanes a year lead to a massive expansion of Pratt & Whitney. By 1943 the company will have nine million square feet of space and employ 40,000 people compared with 3000 in 1938.

1940 Pratt & Whitney begins its first turbine engine research with the PT1. Work ends when the government requires Pratt & Whitney to concentrate on piston engine development and production.

1940 November – General Hap Arnold allows Pratt & Whitney to end all liquid-cooled engine development. The H-3130 had only ground tests beginning in April 1940 and never flew.

1941 April – First run of the R-4360 Wasp Major.

1942 May – Wasp Major first flight.

1942 July – Groundbreaking for massive Kansas City plant to build Double Wasps.

1945 Pratt & Whitney workforce declines to about 26,000.

1945 August – V-J Day. Pratt & Whitey backlog plummets from $400 million to $3 million. Pratt & Whitney and licensees had built 363,619 engines.

1945 October – American Airlines orders 252 new Twin Wasps for DC-4s, the biggest Pratt & Whitney commercial order to that time.

1947 July – Design work begins on the J42 Turbo Wasp after an agreement with Rolls Royce to license Nene engine technology. First run would be in March 1948 and Navy qualification in October 1948.

1947 First test flight of the T34 turboprop on B-17 flying test bed. This was the first axial-flow gas turbine designed by Pratt & Whitney.

1948 Design work on the more powerful J48 based on the Rolls-Royce Tay. First flight would be fall of 1949.

1949 February 26–March 2 – B-50 Lucky Lady II first round-the-world flight using aerial refueling.

1949 May – J57 design start.

# 1940s

1941 June – The U.S. Army Air Corps becomes the U.S. Army Air Forces.

1942 April – Doolittle Tokyo Raid.
September – First B-29 flight.

1943 First flight Lockheed Constellation.

1944 January – Lockheed XP-80 first flight, America's first jet.

1944 September – V-2 rocket, the first ballistic missile.

1946 February – DC-6 first flight.
April – North American begins work on its first rocket engine for the Navaho missile.
August – First Flight Convair B-36.

1947 July – First flight of the Boeing Stratocruiser.

1947 September – U.S. Army Air Forces becomes the independent U.S. Air Force.

1947 October – Chuck Yeager breaks the "sound barrier" in the Bell XS-1.

1947 October – North American XP-86 is America's first swept wing fighter.

1947 November – Howard Hughes, Spruce Goose makes its first and only flight.

1947 December – Boeing XB-47, the first U.S. swept-wing jet bomber, makes first flight.

1949 July – deHavilland Comet first flight.

**1950** January – First run of the J57.

**1951** August – Pratt & Whitney initiates work on nuclear aircraft engine.

**1952** April – First flight of the B-52 protoype. The engine won Luke Hobbs and Pratt & Whitney the 1952 Collier Trophy.

**1953** May – The Pratt-powered North American F100 Super Saber becomes the first production aircraft to exceed Mach 1 in level flight.

**1953** First launch of Rocketdyne-powered Redstone.

**1955** Spring – First flight of the JT4/J75 engine.

**1955** May – Ground broken for CANEL nuclear research facility.

**1956** April – Frederick Rentschler dies.

**1956** July – Experimental JT10 first run, Pratt & Whitney's first turbofan design.

**1957** Fall – First run of the hydrogen-powered 304 engine.

**1957** December – First launch of Rocketdyne-powered Atlas.

**1958** February – JT3D becomes Pratt & Whitney's first commercial turbofan engine.

**1959** May – Florida Research and Development Center formally opens.

**1960** Spring – First run of prototype industrial gas turbine.

**1960** May – First launch of Rocketdyne-powered Delta.

**1961** April – First run of the JT8D.

May – First run of the J58-P-4 engine for SR-71.

**1961** May – First flight test of Pratt & Whitney Canada PT6.

**1961** October – First launch of Rocketdyne-powered Saturn I.

**1964** December – First flight test of TF30, Pratt & Whitney's first afterburning turbofan. First flight of J58-powered SR-71.

**1966** March – First run of the JTF17, Pratt & Whitney's Supersonic Transport engine.

December – First run of JT9D, first Pratt & Whitney high-bypass turbofan

**1967** First run JT15D.

# 1950s

# 1960s

**1952** May – deHavilland Comet enters commercial service with BOAC.

**1954** July – First flight of the 707 prototype, the Boeing Dash 80.

**1956** September – William Boeing dies.

**1957** October – Sputnik launch.

**1957** December – First flight of the Boeing 707.

**1958** January – Explorer I, first U.S. satellite launched.

**1958** May – First flight of the Douglas DC-8.

**1958** October – The National Advisory Committee on Aeronautics (NACA) becomes the National Aeronautics and Space Administration (NASA).

**1958** November – Congress creates the Federal Aviation Administration.

**1959** April – "Original Seven" Mercury astronauts selected.

**1958** Douglas introduces DC-8 .

**1958** F-4 Phantom unveiled by McDonnell-Douglas .

**1958** Pan Am introduces 707 service across Atlantic.

**1959** X-15 makes first flight.

**1961** April – Yuri Gargarin becomes first man in space.

**1961** May – Alan Shephard first American in space.

**1962** February – John Glenn is first astronaut to make an orbital flight; JFK announces Apollo program goal of placing a man on the moon.

**1963** February – First flight 727.

**1965** First flight DC-9.

**1967** April – First flight 737.

**1967** April – McDonnell and Douglas merge.

**1968** December – First flight of the Tu-144, first supersonic transport.

**1969** February – First flight of the Boeing 747.

**1969** March – First flight of the Concorde SST.

**1969** July – Neil Armstrong becomes first man to land on the moon.

**1970** February – JTF22 demonstrator chosen as basis for F-15 engine, the F100.

**1970** December – First flight of the F-14 Tomcat.

**1972** July – First flight F-15/F100.

**1974** February – First flight of the YF-16 prototype.

**1974** August – First run of JT10D.

**1977** March – Flight test of JT8D-200 series prototype.

**1979** October – MD-80 first flight.

**1981** December – First run PW2000.

**1982** February – Development flight testing initiated PW100.

**1983** December – International Aero Engines formed.

**1984** April – First run 94-inch fan PW4000.

**1984** December – Entry into service PW2000.

**1985** December – First run V2500.

**1987** June – EIS 94-inch fan PW4000.

**1989** May – EIS International Aero Engines V2500.

# 1970s

# 1980s

**1970** August – First flight of DC-10.

**1970** December – Airbus Industrie launched.

**1974** December – First flight B-1(B1B October, 1984).

**1975** October – First flight of A300 series.

**1978** November – U.S. airline industry deregulated.

**1978** November – F-18 first flight.

**1980** August – James S. McDonnell dies.

**1981** February – Donald Douglas dies.

**1981** April – First space shuttle flight.

**1981** September – First flight 767.

**1982** February – First flight 757.

**1982** April – First flight A310-200.

**1987** February – First flight A320.

# TIMELINE PRATT & WHITNEY EVENTS & WORLD AVIATION EVENTS

## 1990s

1990 August – PW300 series certification.

1991 April – Pratt -powered YF-22 prototype wins ATF competition.

1992 December – First run F119.

1991 December – PW200 series certified.

1993 August – Certification of the 100-inch fan PW4168.

1994 April – Certification of the 112-inch fan PW4084.

1994 December – EIS PW4168.

1995 May – PW4084 first engine certified for 180-minute ETOPS at entry into service.

1995 June – PW4084 EIS.

1995 December – PW500 series certified.

1996 April – Approval of Pratt & Whitney-NPOEnergomash joint venture for the RD180.

1996 August – Pratt & Whitney and GE form the Engine Alliance to develop engines for superjumbos.

1997 September – First flight F-22/F119.

## 2000s

2000 August – First flight PW6000.

2000 September-October –First flight of Pratt-powered JSF prototypes.

2001 April – First run of Advanced Technology Fan Integrator engine, prototype of the Geared Turbofan.

2001 October – Pratt given formal contract for F135 engines for Joint Strike Fighter.

2002 February – Pratt-led team wins Collier Trophy for JSF lift fan system.

2003 June Pratt & Whitney GDE-1 first flight-weight scramjet to attain Mach 4.5 in tests.

2004 March First run Engine Alliance GP7000.

2004 December – First PW6000 flight on A318.

---

1990 January – First flight MD-11.

1990 April – Hubble Space Telescope launch.

1991 September – First flight C-17.

1991 October – First flight A340.

1992 November – First flight A330.

1993 February – First flight MD-90.

1994 June – First flight 777.

1995 Lockheed and Martin Marietta merge.

1997 Boeing and McDonnell Douglas merge.

1998 First flight MD-95.

---

2000 December – Airbus launches the A380.

2002 January – First flight A318.

2002 August – First flight Eclipse 500 VLJ.

2004 April – Boeing launches 787.

2005 April – First flight Cessna Mustang VLJ.

2005 April – First flight A380.

2005 October – Airbus launches A350 (A350-XWB July 2006).

2007 July – First flight Embraer Phenom VLJ.

**2005** August – Pratt & Whitney completes acquisition of Rocketdyne.

**2005** November – Pratt & Whitney forms new unit, Global Service Partners, to rapidly expand its service business.

**2005** December – Pratt & Whitney Canada certifies initial PW600 for emerging very light jet market.

**2006** November – First GP7000/A380 flight.

**2006** December – First flight F-35 Joint Strike Fighter.

**2007** October – Mitsubishi selects Pratt & Whitney Geared Turbofan™ engine for new regional jet.

**2007** November – First runs of the Geared Turbofan engine.

**2008** February – Bombardier offers C-Series jet with Pratt & Whitney Geared Turbofan™ engine.

# 2000s

The JT8D, Jack Horner's and Bill Gwinn's "best decision."

# Engines That Just
## Kept Going and Going and Going...

## chapter 6

**IT IS SOMEWHAT IRONIC** that what is perhaps the most successful jet engine in Pratt & Whitney history as well as one of the two most produced commercial jet engines in all of aviation started off as a "throwaway."

The JT8D that would power the 727, 737, DC-9, and MD-80 as well as some models of the French Caravelle and the Mercure was based on the J52 turbojet. That engine was originally designed as a one-time-use only engine for the Hound Dog, a kind of early cruise missile. It was simple and rugged, and it turned out would do fine on the A-4 Skyhawk and A-6 Intruder Navy attack aircraft. It had a production run from the early 1950s until 1986.

So how did a throwaway become the hugely successful JT8D? Well, it started with a phone call from Charles Froesch of Eastern Airlines to his friend Frank DuLyn at Pratt & Whitney in early 1960. Eastern was one of the original customers for a new jet from Boeing, the 727. This would be the first plane that would get jet travel down to short- and medium-range routes, replacing Convair 440s, Martin 2-0-2s, and even the many DC-3s still flying. It was unique for its time in that it would have three engines, all mounted at the rear. When Boeing began 727 development in 1959, the Rolls-Royce Spey was specified. It quickly became apparent that the 727, like seemingly all new aircraft, was getting heavier, and there were doubts the Spey could grow. Hence the phone call from Froesch to DuLyn. Did Pratt have anything that would work on the 727?

At the time Pratt & Whitney had been developing a very advanced engine known inside the company as the JTF10A. It was based on the core of the TF30 fighter engine, but aircraft companies and airlines were wary of its advanced technology in a commercial application. If Pratt were going to get on the 727, and Boeing was not excited about that prospect in the middle of a development program, it had to come up with something clearly superior to the Spey, rugged enough for short-range commercial service and an engine that could be developed quickly. The J52 was about the right size at 11,200 pounds of thrust,

The JT8D and the JT8D-200 would become the most-produced jet engines in commercial history in their time; many will fly well into the 21st century.

but it was a military turbojet totally unsuited for commercial service. Once again Pratt people went on one of those 24-hours-a-day, seven-days-a-week crash programs and came up with the JT8D. The paper concept design was done over exactly one week in March 1960. It would use the J52 core with a new low-pressure section and a two-stage fan. For the center engine to fit into the tail installation, it would have a full-length duct for by-pass air. The more typical arrangement was for bypass air to exit the engine well forward on the engine nacelle. This would not work on the 727 with its center engine buried inside the tail and aft fuselage.

With the design concept in hand, Pratt & Whitney made the decision to go ahead with development, $75 million worth, a lot of money in 1960. A few years later Jack Horner said, "It was the best decision Bill (Gwinn) and I ever made." First run of the engine was April 7, 1961, and it entered service in 1963. Pratt & Whitney would build nearly 15,000 JT8Ds before serial production ended in 1996.

The 727 was a huge success, but Douglas saw another opportunity for a somewhat smaller airplane. That would become the DC-9, also powered by the JT8D. Boeing would respond to the DC-9 with the original 737, and once again the JT8D was the engine choice. Those three aircraft and the JT8D simply overpowered competition from the Trident, the BAC-111, and the Caravelle. Airlines loved the JT8D. It was simple, extremely reliable, easy, and relatively cheap to maintain and operate.

The JT8D even had one military spin-off as the Volvo RM8 for the Swedish Viggen fighter.

By the late 1970s Douglas was looking for a way to extend the life of its DC-9 line. It needed a somewhat bigger, more fuel-efficient aircraft. Pratt & Whitney redesigned the original JT8D into the JT8D-200 series. The thrust was increased from 14,000–17,000 pounds to 20,000–21,700 pounds with a new fan and other modifications. More than 2000 JT8D-200s would be built for the now McDonnell-Douglas MD-80.

By the 1990s new, stringent noise regulations seemed to spell the end of the JT8D. Pratt & Whitney, working with companies like Nordam, tackled that issue too. Hush kits were developed that would bring 727s, 737s, and DC-9s into compliance with Stage III noise rules. Several thousand JT8Ds will probably fly well into the 21st century, and yet another application for the engine appeared. The JT8D-200 was specified for the Air Force's re-engining and modernization program for the 707-based JSTARS aircraft.

Not a bad run for a throwaway engine that was designed in a week's time back in 1960.

Photo courtesy The Boeing Company.

The JT8D-powered 727.

# A REALLY, REALLY BIG FAN

By the early 1960s the U.S. Air Force was wrestling with the issue of airlift. Planes like the C-130 and C-141 were great, but did not have the capacity the Air Force needed to move really big cargos long distances. For instance, the C-141 actually could carry far more gross weight than it could accommodate in terms of space. By 1964 a series of

" It was the best decision Bill (Gwinn) and I ever made. "

initial studies resulted in a competition among Boeing, Lockheed, and Douglas for a new very large airlifter that would ultimately become the C-5A. Pratt & Whitney and GE got contracts to develop engines for the massive new airplane.

Both companies recognized that the only way to meet the requirements of the C-5A was to expand the whole turbofan concept significantly. Turbofans were efficient by moving some of the air taken into the engine around the engine's core instead of through it, the so-called bypass ratio. Simply put, it is more efficient to accelerate a large amount of bypass air moderately than to accelerate a smaller amount of air greatly down the core of the engine. Engines like the JT3D and JT8D had bypass ratios ranging between 1.1 and 1.4,-to-1 most of the air still going down the core. To get the much higher thrust that would be required for the C-5 while

The JT9D represented a step change in engine size and power and needed massive new test cells at the Middletown, Connecticut, plant.

keeping fuel burn reasonable, a much higher bypass ratio would be required. As much as four or five times as much air would have to bypass the engine core as go through it. This meant a big fan at the front of the engine to grab all of that air.

GE had been working on high-bypass concepts for some time and won the C-5 competition in August 1965 with the TF39, a huge win with ramifications felt to this day in the engine business. The TF39 would become the very successful CF6 series of GE commercial engines. Subsequently, of course, Lockheed won the airframe competition for the C-5.

But the two "losers" were undaunted. Maybe you could take the designs and come up with a commercial aircraft. The Lockheed airplane was not very suitable for commercial work given its high wing mounting. The Boeing design was much more workable, and the Pratt & Whitney engine, the JT9D, would fit it perfectly.

In England, Rolls-Royce was also working on high-bypass designs. It also focused much of its effort on engines for the supersonic Concorde, a technological marvel that was a financial disaster. The company was also having some difficulty because its two big commercial programs, the Trident and the BAC-111, were not huge successes compared with the cash-generating JT8D. Rolls would do its best to get into the high-bypass race with the RB-211 family that found its first home on the Lockheed L-1011 Tristar. Unfortunately, the L-1011 was not a success, and Rolls had serious troubles with the original composite fan for the engine. It wound up in bankruptcy and was taken over by the British government, limiting its ability to compete for a time. Rolls, however, would come roaring back.

Meanwhile, Boeing and Pratt & Whitney were searching for a customer for their aircraft/engine offering, the 747/JT9D. Not surprisingly, Juan Trippe, the founder of Pan Am, stepped up after flirting with both Lockheed and Douglas. He drove a very hard bargain with lots of performance guarantees. Boeing was betting the company and so was Pratt to some degree. Lawrence Kuter was a Pan Am vice president who offered that Boeing, Pratt, and Pan Am would have to push the state of the art in virtually every phase of design, manufacturing, and airline operations to make the plane a success. He dubbed the 747 "The Great Gamble."

For Boeing it meant building the monstrous assembly plant at Everett, Washington, the largest building in the world under one roof at the time. For Pratt & Whitney it meant building new assembly and test facilities at the old CANEL property in Middletown, Connecticut. The East Hartford plant just did not have the high bay areas and large enough test cells for the JT9D, which was much larger than anything Pratt & Whitney had built to that time.

The JT9D program would have more than its share of heartache for all involved and would strain the long and close relationship between Boeing and Pratt & Whitney. As in the past, Pratt & Whitney would rise to the tasks but not without

The contrast between the JT9D (left) and the JT3D (right).

(apologies to Winston Churchill) copious amounts of sweat, tears, and some would say blood.

As always, the issue was weight. Pratt's original concept for the JT9D for the C-5A was a 36,000-pound thrust engine. As the 747 took shape, that was boosted to a 41,000-pound requirement, and then there was talk that the engine would need 43,000 pounds or even 44,000 pounds of takeoff thrust.

The 747 kept getting bigger and heavier by as much as 100,000 pounds over the development, and the JT9D was breathing very heavily to get the airplane in the air and meet range, payload, and fuel burn guarantees. Both Boeing and Pratt & Whitney were taking aviation into a whole new realm with the size and power of the 747/JT9D. Some Pratt engineers of that day say

privately neither they nor Boeing fully appreciated that jumping from the 707/JT3D to the 747/JT9D was not just arithmetic—three times larger, more powerful, etc. It was geometric, orders of magnitude different.

Aviation buffs around the world love to tell stories of this time period; 747s sitting at Boeing with no engines, and big concrete blocks tied to the wings. There was the embarrassment for Bill Gwinn, a passenger on the first 747 flight in 1970, when the engines would not start properly.[1]

Pilots had to push the JT9Ds hard on the 747. Higher temperatures wore out parts faster and hurt fuel burn. Also engine cases would "ovalize," that is, get out of dimension. This caused leaks in the engine flow that hurt efficiency and thrust. The engine was not an initial success by any means, but Pratt people tackled the issues one by one and followed the old principle that work was not done until one got the last ounce of performance and reliability out of an engine. The JT9D would grow from a 43,000-pound thrust engine into the JT9D-7R4 series at 56,000 pounds of thrust. JT9Ds, although out of production since

Together the JT9D and the 747 changed commercial aviation in the 1970s.

1990, still fly on 747s, 767s, DC-10s, A300s, and A310s and will for years to come.

Douglas came out with the DC-10 as a competitor to the 747 on thinner routes and

---

[1]Gwinn had taken over from Horner as UAC chairman on October 1, 1969. September 30 had been the rollout of the first 747. Horner later recalled that at the dinner that night in Seattle he realized at 9:00 p.m. that it was 12:00 a.m. October 1 in East Hartford: "So I walked across the floor and got a hold of Bill and stood him up and said, 'Well, Bill, here it is, take it. You are now chairman and chief executive officer.'"

Photo copyright Airbus. Used with permission.

in 1968 chose the GE CF6. This was the commercial version of the TF39 that had won the C-5 competition. It was a major inroad for GE and led to the CF6 getting on the 747 in 1972 after the JT9Ds difficulties were well known. Rolls-Royce, beginning its comeback, would win a place on the 747 three years later. The day of single-source engines for commercial airplanes was over.[2]

# WHERE IS TOULOUSE?

By the late 1960s the European aerospace industry had to be discouraged. Its military budgets were small relative to the United States, and in some cases European governments had to buy U.S. equipment to stay ahead of Soviet gear. European commercial programs had been dismal failures for the most part. European leaders gradually came to realize that there would never be success as long as there were British programs, French programs, German programs, and so on. Europe had to unite to take on the Americans with any chance of winning. So in 1969 Airbus Industries was

---

[2] It was perhaps the DC-10 that had started all this. After GE won the initial application, Northwest Airlines said they would not buy DC-10s unless they had Pratt engines. So Pratt stepped up with a modification of the 747 engine. Industry observers at the time believed that angered Boeing and pushed the effort to get GE and later Rolls on the 747. Only Northwest and later JAL bought Pratt-powered DC-10s.

created. It would build a widebody that would fit in just below the DC-10 and L-1011, and it would be a twin-engine aircraft. But whose engines would they be?

Rolls-Royce had a fairly strong relationship with the French national engine builder, SNECMA. In fact, Rolls had the edge on powering Airbus's first offering, the A300. But the British company was also working on the RB-211 for the Lockheed L-1011 at the time. It did not have the resources to tackle both programs, and Lockheed, a well-established American company, perhaps looked like a better bet than this new outfit headquartered in Toulouse, France.

Pratt & Whitney also had been working with SNECMA for many years and in fact in 1959 had taken a 10 percent interest in the French company. SNECMA and Pratt worked together on some military programs around the TF30 fighter engine, but the relationship never fully blossomed. Even with SNECMA urging Pratt to get involved in this new Airbus program, there was not a great deal of interest in East Hartford. This was perhaps because so many resources were focused on the 747 and DC-10 programs and new military business in the United States. And it might be hard to recall today with the outstanding success of Airbus that in the early 1970s many, many people believed the consortium would never even build an airplane let alone become a market force against Boeing and Douglas.

GE was in a somewhat different position. It had had no commercial business to speak of for many years and was just getting going on the DC-10. The 747 was still a few years away for GE. So here was an opportunity. GE struck a deal with SNECMA that, as we will see, had far-ranging implications. And GE became the launch engine for the A300 and the A310.

Pratt & Whitney fought back and won places on both airplanes with the JT9D-7 series, but it would take time to build up a strong relationship with Airbus.

# "Little Pratt"
## Does Big Things

## chapter 7

**PRATT & WHITNEY CANADA** could be described as the little company that could. From a small service shop it has grown over the years into a worldwide organization with more than 42,000 engines in service with 9000 operators in 190 countries. Often referred to as "Little Pratt" both inside and outside the company, the slang title really no longer fits and has not for many years.

Back in 1928, however, that name might have been apt. "Big Pratt" saw opportunity in the emerging Canadian market but needed a way to enter it. There was still a British Empire in those days, and its tariff system favored British engines in Canada. Engine parts were taxed at a lower rate than complete engines, and so Big Pratt could ship kits to Canada where Little Pratt would put them together. The Canadian company would also offer engine service to Canadian operators on their doorstep with a unique understanding of those customers' requirements. So Canadian Pratt & Whitney was formally incorporated in late 1928 and opened its doors in the Montreal suburb of Longueuil in February 1929 with 10 employees.

James Young was the company's first president. He had previously worked for Niles-Bement-Pond and Pratt & Whitney Machine Tool. Considered one of the early outstanding leaders of Canadian aviation, Young would serve as Pratt & Whitney Canada president until 1947 and remained on the board of directors until 1964. Pratt & Whitney never intended its Canadian sibling to become a full-up engine manufacturer. It would assemble engines destined for the Canadian market and provide overhaul and repair

Pratt & Whitney Canada opened its doors in Longueuil, Quebec, in early 1929 with 10 employees to assemble, overhaul, and service Wasp and Hornet engines.

services. The leaders of the Canadian company, however, always held the thought in the back of their minds that they would get into actually making things one day.

The service business would be the heart of the operation for many years, and it taught Pratt & Whitney Canada valuable lessons. Many of its customers were small operators flying utility aircraft in the wilds of Canada, the famous bush pilots so romanticized today. The Canadian company quickly learned how to provide top-notch service to very small operators, often located not just in the middle of nowhere, but on the other side of nowhere. It was a lesson that has stood the company in good stead even today when it supports 42,000 engines on 23,000 aircraft scattered among 9000 operators in 190 countries through a network of some 30 service facilities on six continents.

# HUSTLING BUSINESS

Because Pratt & Whitney Canada was somewhat limited in its scope, its people were always scrambling for a bit of new business here and there. For instance, by 1930 Young had negotiated a deal with Hamilton Standard to service its propellers in Canada. That would be the seed of Pratt & Whitney Canada's first real manufacturing operation.

With the coming of World War II, Pratt & Whitney Canada like every aviation company had to grow almost exponentially. In 1939 the company had 17 employees and did $488,000 in business. By 1943 there were 440 people at work, and sales had climbed to $8.5 million. Pratt & Whitney Canada concentrated its efforts on the Wasp engines that were used in training/utility aircraft such as the Harvard and the Norseman. Again, its role was limited to assembling U.S-built engines for the Canadian airframe manufacturers. The Canadian government had pretty well set a policy that the best way for Canadian manufacturers to contribute to the war effort was to build proven U.S.- or British-designed products. In 1941 this led Pratt & Whitney Canada to begin its first full-up manufacturing operation, Canadian Propellers Ltd. The company was set up to produce the two-bladed Hamilton Standard 12D40 used on light aircraft. It showed the aviation industry that Pratt & Whitney Canada could manufacture equipment, not just assemble and service it.

John Drummond was the tooling and production engineer for Canadian Propellers and recalled those days in the company history, *Power: The Pratt & Whitney Canada Story*:

> We started turning out 12D40s weeks ahead of schedule and the government asked us to concentrate on that. We reached a peak of over 1000 12D40s a month. The efficiency of our plant and the quality of our propellers was soon recognized as second to none. We shipped propellers out by the carload to Canadian and U.S. destinations.[1]

---

[1] For those interested in a more detailed history of Pratt & Whitney Canada, the author recommends *Power, The Pratt & Whitney Canada Story* by Kenneth Sullivan and Larry Milberry, 1989 CANAV Books, Toronto, an invaluable resource for this writer.

Another key growth step for Pratt & Whitney Canada came in its support of Ferry Command, the group that had to get thousands of airplanes from Canadian and U.S. factories across the Atlantic. Crossing "the pond" was still a pretty big deal in 1939, usually only done by flying boats or airships up until that time. The Pratt & Whitney Canada team learned how to support a flood of engines and propellers that passed through their hands on the way to the European battlefront. This experience would only strengthen Pratt & Whitney Canada's outstanding support capability developed in the years of bush flying.

With the end of the war, Pratt & Whitney Canada like every other aviation company had to come up with some way of staying in business. Perhaps because it had always been a small organization, Pratt & Whitney Canada had a true entrepreneurial spirit. It would go into the surplus business. John Drummond recalled:

> Only the Wasp and Wasp Juniors were widely used in commercial aviation in the North. They had been produced since the war mainly by licensees. We also knew that thousands of wartime aircraft would soon be on the commercial market. Their operators would need a reliable source of engines and spare parts. . . . Hartford took the view it didn't want anything to do with the market. It might affect their reputation if they took responsibility for a licensee's product. They also felt the availability of such products would interfere with the sale of new engines and parts. I took a different stand—that we had to face the fact that surplus products were a reality; they were going to be used. They would need support, and if we didn't provide it, some other company would.

James Young had to go to Hartford and do some smooth talking, but Big Pratt and United Aircraft agreed, and Pratt & Whitney Canada went into the war surplus business. Again, it was forging a link in the chain that would lead to its becoming a major manufacturer. Around the same time Pratt & Whitney Canada became involved in an entirely new kind of business—helicopters. It became the Canadian agent for UAC sister company, Sikorsky. Pratt & Whitney Canada people learned about a whole new aviation segment, again building a knowledge base when the company began building engines for helicopters, one of its key markets in years to come.

As has been noted, the people of Pratt & Whitney Canada had always had an eye on becoming a true manufacturer, and by the late 1940s it was becoming a necessity. Ron Riley, who had taken over as president in 1948, said, "If the company was to continue its growth and development, it must change the character of its business and commence manufacturing." The Korean War brought that opportunity. Big Pratt was being flooded with orders for J48s and needed help. The original idea was for Pratt & Whitney Canada to get into manufacturing J48 parts, but the Canadian government encouraged the folks in Longueuil to get involved in building a complete engine, the Wasp. It was estimated that there would be demand for lots of the ubiquitous engine.

In 1956, the first steps were taken to form a design team of gas turbine specialists to lead the development of Canada's first small turbine engine, the PT6.

John Drummond recalled that the requirement was 250 engines a year for the U.S. market and 50 for Canada. "We put together a proposal. The government accepted it and asked us to set aside the J48 (parts) project. This we did, for we now had a contract that would make us a major manufacturer." In late 1950 a U.S.– Canadian agreement was reached waiving many of the "Buy American" restrictions on defense-related products, and in 1951 a formal contract was signed for Pratt & Whitney Canada to produce 1000 engines plus spares over the next five years. After the end of the Korean War, the demand for Wasps fell, but by then Pratt & Whitney Canada had worked out an agreement with Big Pratt to become the source for all piston engine spare parts.

Employees from those days remember journeying frequently to Hartford to learn the job of building engines. The preferred mode of travel (at least by the company travel office) was the Washingtonian; also known as the Night Train to Hartford. It left Montreal about 8:30 p.m. and usually got in behind schedule early the next morning. The trip back left around midnight, when it was on time, and got back to Montreal around 9:00 a.m. "These were the years when many a father who worked at Pratt & Whitney Canada was known to his children as a voice over the telephone," according to the account in *Power: The Pratt & Whitney Canada Story*.

One could say the Wasp work was one of the most critical steps in Pratt & Whitney Canada's history. Veteran company man Bob Losch summed it up succinctly: "We were responsible for the cost and selling of those parts, and that entitled us to keep the profits. With those profits we later developed the PT6."

# "FROM LITTLE ACORNS GROW . . ."

If Pratt & Whitney was "made" by the Wasp, then Pratt & Whitney Canada was made by the PT6. Over the years some 90 different models of the engine have been customized for the widely varying demands of customers. Horsepower ranges from 580 to over 1900 and over 43,000 PT6s have been built. It is clearly the most popular light- to medium-duty turboprop/turboshaft in the world. Those profits from the Wasp had, indeed, been invested wisely.

Actually, the PT6 was not the first gas turbine Pratt & Whitney Canada pursued. In 1957 the Canadian government was looking for a new trainer, and Pratt & Whitney Canada started work on a small turbojet known as the DS-3J and then the DS-4J. Development work started, but the challenge seemed a bit more than Pratt & Whitney Canada was ready to tackle, financially or otherwise. So the program was turned over to East Hartford, where it became the JT12. The people in Longueuil were disappointed. But, in the long run, it might have been a blessing in disguise because they could now focus on what would become the PT6.

They had been doing a lot of market research during 1958 on what was available or on the drawing boards. Teams went to Piper, Beech, and Cessna. The idea was to find a market spot that was relatively open for a new engine. Allison was working on engines in the 250-shaft horsepower range. Rolls Royce had the Dart in the 2000-shaft horsepower range, and Allison was dominating the big turboprops with its T56 family. In the end the team led by Kenneth Sullivan and Elvie Smith (later president of the company) recommended that Pratt & Whitney Canada develop a 450-horsepower turboprop that could grow to 500 shaft horsepower, a range that would fit Canadian aircraft like the Otter and the Beaver. The goal was to build an engine with the same operating costs as small piston engines, but that pushed an aircraft 50 miles an hour faster.

> These were the years when a father who worked at Pratt & Whitney Canada was known to his children as a voice over the telephone.

Then the engineering team looked at what kind of engine to build. It was decided that a free turbine engine was a better bet than a fixed-shaft engine. Kenneth Sullivan and Larry Milberry explained the reasoning in their book, *Power, The Pratt & Whitney Canada Story*:

On the fixed-shaft engine, the gas generator and power turbine share a common shaft. On the free turbine there are two units, one driving the compressor and one producing the power. The link between the two is not mechanical but is made by the flow of hot gases through the engine. The fixed shaft engine requires fewer parts, so is cheaper to develop. The free-turbine is more complex, hence costlier, but has such advantages as requiring less starting power and simpler fuel controls. The free turbine eliminates clutch requirements in a helicopter and makes easier the pairing of engines for more powerful installations. Fixed-wing aircraft could use an off-the-shelf propeller with a free turbine instead of a costly tailor-made one required by a fixed-shaft engine.

Airplane people liked the free turbine approach because in an engine-out situation it produced less drag. The windmilling propeller would only have to turn the free turbine, not the whole compressor and turbine. The aircraft structure need not be as beefy and thus heavier as in fixed-shaft setup.

Looking back from today, it is easy to see how simple this all was, how logical, how easy. It most definitely was not at the time. On December 1, 1958, the Canadian team went to Hartford to make their case to Wright Parkins, Big Pratt's top engineer. A sort of "red team" in Hartford had done a design as well. Parkins looked at both and gave the Canadian the go-ahead. But it would be five years before a production engine emerged, and there would be a lot of very hard work along the way.

Pratt & Whitney Canada veteran Allan Newland recalled later that the inexperience of the Pratt & Whitney Canada team showed: "We had no history, no experience as a team. This was a far cry from what would happen in a mature organization with a long history of design. Our inexperience did, however, have a positive aspect—we were uninhibited. We had no past failures."

By February 1960 a complete PT6 with a propeller went on test, but there was a host of technical problems, and costs were getting very high. Parkins down in East Hartford was monitoring the program and decided that "your engineers are very intelligent, but are inexperienced and don't understand engine development." A high-power meeting of all of the top brass finally decided a SWAT

Some 90 different models of the PT6 engine have been developed.

team was needed, and in early 1961 technical direction of the PT6 was taken over by a six-man team from Big Pratt led by the redoubtable Bruce Torrell. A native of Winnipeg, Torrell had been around the engine business a good deal with stints at Canada's National Research Council and even worked for a time with Sir Frank Whittle's Power Jets in the United Kingdom. He had come to Big Pratt in 1946.

> " He was known to show up in the middle of the night wearing a raincoat over his pajamas. "

Elvie Smith summed it up: "We learned how to develop engines from Bruce Torrell." More than a little the task master, Torrell ran the program with an iron hand, scrapped the one-shift-a-day approach and went to around-the-clock development testing. One of the people working on the engine, Rick Stamm, recalled: "When he was in town, he could be found in the plant at all hours. He was known to show up in the middle of the night wearing a raincoat over his pajamas."

If Torrell made the PT6, it might be that the PT6 in some way made Bruce Torrell. He became Pratt & Whitney president in 1971.[2]

# IN THE AIR

By the summer of 1961, the PT6 was flying aboard a Royal Canadian Air Force Beech Expeditor, the engine mounted in the nose between the two Pratt piston engines. It was said at the time that air traffic controllers were amazed when the test pilots took the aircraft up to 26,000 feet, something the Expeditor was never designed to do on just piston power.

The first aircraft to fly solely on PT6 power was actually a helicopter, the Hiller Ten99 prototype in July 1961. Piasecki, Lockheed, and Kaman also all used PT6s during this period in various prototypes. There were also projects to convert piston-powered DeHavilland Otters and Beavers to the PT6. Although technically successful, most of the small bush operators could not afford the conversion.

The big breakthrough for the PT6 came from Beech. The Wichita-based company had long used Wasps and Wasp Juniors and was now cautiously looking at turbine power for a new Army liaison aircraft based on its piston Queen Air. The prototype was known as the NU-8F. It first flew in May 1963.[3] Army testing

---

[2] Thor Stephenson, who served as company president from 1959 to 1975, said later that the PT6 almost died before it was born: "The early days of the PT6 program were not encouraging, technically or sales-wise. As a result, James Young, Pratt & Whitney Canada founder, and his great friend on the board, Hubert Welsford, went to Hartford to see Horner (UAC Chairman Jack Horner.) They wanted the PT6 terminated and Pratt & Whitney Canada to revert to a sales and service organization. Horner rejected their pleas and the PT6 continued."

[3] Company lore says it flew, sort of, one day earlier than scheduled. During high-speed ground tests, so the story goes, the aircraft hit a bump on the runway and was airborne for a short time.

went well, and Beech then evolved the aircraft, modifying a piston-powered Queen Air design into the turbine-powered King Air. This landmark airplane flew on January 20, 1964, which happened to be the 28th anniversary of the first flight of the Beech 18. But would the aircraft sell? The fourth King Air went on a European tour in the fall, and in three weeks 27 were sold to companies like Volkswagen and Daimler-Benz. Even the Aga Khan bought one. The PT6 was off and running.

The next big break came with DeHavilland's decision to build the turbine-powered Twin Otter. It became a marvelous utility aircraft all over the world and could perhaps be seen as the first commuter airliner. DeHavilland would build more than 800 Twin Otters. The PT6 became a favorite for military primary trainers after its installation on the Beech T-34 for the U.S. Navy as well as several international customers. One disappointment in those years was the COIN (counterinsurgency operations) program. Convair picked the PT6 for its Model 48 Charger, but the competition was won by the North American OV-10 Bronco powered by AiResearch Garrett TPE331s. Garrett (now Honeywell) was and still is a major Pratt & Whitney Canada competitor.

The PT6 also helped Pratt & Whitney get into a whole new business: auxiliary power units. Working with Hamilton Standard, the PT6 was adapted as the APU for the Lockheed L-1011. The little engine seemed adaptable to almost anything. It was used in marine and industrial applications, including a giant snowblower for the British Columbia government. And it even powered a train. This was the Turbo Train, an experimental high-speed train that was developed by United Aircraft's Corporate Systems Center. The idea was to meld various technologies in the corporation's separate divisions, which usually worked pretty much autonomously. Sikorsky and Pratt & Whitney Canada worked on the Turbo Train that could run up to 170 miles an hour with its four ST6B industrial engines. The experimental trains ran in the United States and Canada between 1968 and 1982 but never caught on as an alternative to diesel-electric. Part of the problem was that the roadbeds the trains ran on could not handle the speeds it was capable of. But today the ST6 engine is back in the railroad business powering new experimental high-speed trains. Time will tell if the effort will be crowned with success this time.

# THE RACER'S EDGE

People of a certain age might remember television commercials from the 1960s for the fuel additive STP that featured trench-coated company president Andy Granatelli and used the catch phrase, "The Racer's Edge." Granatelli had sponsored Indianapolis 500 teams for several years as a way to promote his company. In 1967 he and Pratt & Whitney Canada combined for one of the most unusual applications of any Pratt & Whitney engine, a PT6 powering an Indy racecar.

Observers at "The Brickyard" were amazed when the STP special came onto the track and whisked around almost noiselessly compared to the deep-throated roar of Offenhauser-powered traditional cars. It was instantly dubbed the "whooshmobile." The United States Auto Club (USAC), the sponsor of the race, tried to kick out the STP car, ruling that the car had to use its sponsor's product. It turned out STP was a perfect lubricant for the PT6 fuel pump. The car looked like a winner for the first 492 miles of the race under the steady hands of driver Parnelli Jones. Then a $6 bearing in the transmission failed, and the car glided to a stop. Granatelli came back in 1968 with five turbine cars. Once again USAC attacked, limiting the size of air intakes to 16 square inches. The PT6 needed 22. Pratt & Whitney Canada field personnel modified the engine to meet the requirement. Three cars qualified for the 500, but small mechanical failures killed their chances, none related to the production PT6. USAC then instituted even more air

Pratt & Whitney Canada's Boeing 720 Flying Test Bed has played a key role in the development and certification of a wide variety of engines for more than 20 years, including the PW150A turboprop engine shown here.

intake rules, and turbine cars were out for good. But Andy Granatelli loved the PT6 and had one specially modified for his Corvette.

Today the PT6 family is still going strong, certified on over 130 fixed-wing and helicopter models. And the engine seems to find new applications all of the time such as an improved King Air and a Socata TBM 850. Some 35,000 turboprop engines have been built and more than 8000 turboshaft single and twin-pack engines. Combined turboprop and turboshaft hours are over 300 million for some 6000 operators worldwide.

So Pratt & Whitney Canada had at last become a manufacturer as well as one of the most respected service organizations in aviation. The little engine that Big Pratt engineers jokingly said they could tuck under their arm compared with the behemoths in East Hartford was the foundation that the people of Pratt & Whitney Canada would build on.

# TURBOPROPS TO TURBOFANS

Pratt & Whitney Canada had the success that it had yearned for, but it did not want to become a one-product company. Once again the people in Longueuil looked around the market for a spot to enter a new market. The opportunity came from Wichita again. Cessna was losing business to its neighbor Beech because the King Air was clearly superior to piston-powered corporate aircraft. But many customers did not want to move all the way up to an expensive jet-powered aircraft. Cessna came up with the concept of a simple, relatively low-cost turbo-fan-powered plane that could fly at 400 miles an hour. The straight-wing design would handle so easily that piston pilots could move up to it easily.

Cessna knew about Pratt's emergence as an engine designer and builder and asked what could be done. In June 1966 detailed design began on what would become the JT15D, a 2000-pound thrust turbofan. The engine had a single-stage centrifugal compressor, a technology Pratt & Whitney Canada knew intimately. The turbine was redesigned from a single stage to dual stage to get the engine diameter down to fit the Cessna design known as the FanJet500. The development went well, and in September 1968 Cessna ordered 50 JT15Ds for what was now called the Citation. Once again Pratt & Whitney Canada had a winner. Since entering service in 1971, the JT15D has been constantly upgraded and improved. It can produce up to 3350 pounds of thrust. Besides the many Cessna models, the engine found homes on the Beechjet 400 and the T-1A Jayhawk. The JT15D was the genesis of a family of turbofan engines, much as the PT6 was the beginning of a family of turboprop and turboshaft engines that would greatly extend the reach of Pratt & Whitney Canada. It was not a one-product company anymore.

# BIGGER AND BETTER

As the 1970s moved forward, Pratt & Whitney Canada started looking at a new market, turboprops bigger than the PT6, something in the range of 1500–2500 shaft horsepower. The regional/commuter market was beginning to grow, and the company had already boosted horsepower on the PT6 to fit new applications like the Shorts 330 and the DeHavilland Dash 7.

In 1978 the United States deregulated the airline industry, and the commuter/regional market really took off. Regional airlines could now fly 60-passenger airplanes rather than the 30-passenger Twin Otters and Beech 99s powered by the PT6. Pratt & Whitney Canada wanted to, if you will, succeed itself in the regional/commuter market. The timing turned out pretty well because by 1981 or so the corporate market, a big chunk of the PT6 and JT15D business, was sagging badly.

During this period, Pratt & Whitney Canada came up with an approach to engine development that has since migrated throughout the company, technology demonstration engines (TDE). The idea is to build a prototype engine that pretty much fits into the market slot the company wants to go for as quickly and inexpensively as it can. The engine does not have to be exact as long as it can get solid technical answers. The "proof-of-concept" engine speeds up the development of an ultimate product and cuts the cost. By 1978 Pratt & Whitney Canada was running TDE-1. By 1979 that learning process kicked off full-scale development of the first PW100 model of turboprops. The competitors would be an improved Rolls-Royce Dart, a reliable, but very old engine by then, and the GE CT7. Pratt & Whitney Canada almost simultaneously with the original development began work on a larger model to beat the competition.

As engine development got underway, there were several candidate airplanes. A lot of companies were getting into the commuter/regional market. Talking with these companies, Pratt & Whitney Canada engineers came to the conclusion that they would have to boost horsepower to a range of 1800 to 2600 to meet the varying requirements. Work was also done with sister company Hamilton Standard on a new composite propeller for the new generation of regional airplanes. Once again Pratt & Whitney Canada, with market savvy and careful engineering, had come up with a solid product. The design life of the turbomachinery was 30,000 cycles. (A takeoff, cruise, and landing is one cycle.) The engine had only six major rotating components on a three-shaft, two-spool layout. The experience gained on the PT6 and the JT15D was instrumental in the design of the PW100 as a free turbine engine. The engine modules could be mixed and matched to give a wide range of power for various applications, and there was room for growth. In fact, today the PW150 model can produce 5000 shaft horsepower.

Another factor in the success of the new turboprop was Pratt & Whitney Canada's reputation among regional/commuter airlines and aircraft companies. The PT6 had shown them that the people in Longueuil understood what their customers needed and gave excellent support. In 1979 Embraer chose the PW100 for its EMB-120 commuter. DeHavilland picked the engine for its new Dash 8. Fokker would pick it for its F-50 and F-60 as would British Aerospace for the Advanced Turbo Prop (ATP). Another major win was for the Franco-Italian ATR 42 that would grow into the ATR72. GE would take the CASA CN235 and the Saab 350, but by the mid-1980s Pratt & Whitney Canada was the clear leader in commuter/regional turboprops.

That leadership would be challenged mightily very soon by the emergence of regional jets. By the late 1980s there was a lot of talk on the need for bigger, faster regional aircraft. There was even talk that the regionals would be looking for turbofan planes. Pratt & Whitney Canada was skeptical because these would be big, expensive airplanes that would not have the fuel efficiency of the turbo-prop. But the regional market was booming, and the bigger planes came. Allison, soon to become part of Rolls-Royce, came up with a whole new family of engines, the AE2100 turboprop and the AE3007 turbofan. General Electric developed the CF34 turbofan. Pratt & Whitney Canada really had no matching engine. It seemed the ship had sailed with Pratt & Whitney Canada still on the dock. The response, however, was a greatly improved PW100 engine, the PW150, that became the powerplant for the Dash 8-Q400. And, lo and behold, the turboprop market started making a big comeback not too long after the turn of the century. Fuel prices started to soar, and turboprops looked mighty attractive again. The PW100 family with 6100 engines in service on 2200 aircraft looks like it will have a long run.

And Pratt & Whitney Canada is doing the technology homework for the new PW800 family, a turbofan that will range between 10,000 and 20,000 pounds of thrust. Already the company has run a technology demonstration engine that has shown the PW800 design greatly increases fuel efficiency and reduces noise. The engine bridges the gap between the PW300 series of 5000- to 8000-pound thrust business jet engines and the 20,000-pound thrust PW6000 from East Hartford that flies on the Airbus A318. The engine targets new regional aircraft and large business jets.

# A LOT OF ENGINES, A REAL LOT

With the confidence and technology built through solidly successful engine programs, Pratt & Whitney Canada was off and running with more new engine families. The PW200 turboshaft series went after new-generation light twin-engine helicopters such as the Bell 429, the Agusta Grand, and the Sikorsky S-76D, building off the market success of the PT6 turboshaft engines. Two new families of turbofan engines emerged to follow on the JT15D. The PW300 was the next step at 4500 to 8000 pounds of thrust. The PW500 series ranges in

Launched in 2003, the PW600 engine family is leading the way in the very light jet market.

thrust from 2700 to 4500 pounds. Cessna uses the engine on a variety of derivative Citation models. And Embraer picked the PW500 for its Phenom 300 bizjet. The PW900 family of auxiliary power units is used on the 747–400 and the A380.

But as the 21st century opened, perhaps the development being watched most eagerly was the PW600 family for the very light jet market (VLJ). VLJs made Pratt & Whitney Canada think about designing and building engines in a whole new way. The new aircraft would have to be inexpensive, easy to fly with one pilot, and easy to maintain. As for manufacturing, the numbers could be stunning, well over 1000 engines a year, like nothing seen since World War II. Manufacturers forecast thousands of VLJs, especially if the air taxi market developed as many entrepreneurs envisioned. Engine builders would have to produce at almost unheard of rates at very low costs if the whole concept was going to work. The catch phrase "step-change" became part of the vocabulary in Longueuil.

But first Pratt & Whitney Canada had to come up with an engine. In 1999 it went to its tried-and-true system of a technology demonstrator engine and from that emerged the PW600 family. Eventually there would be three models ranging from 900 pounds of thrust up to 3000 pounds. The engine would weigh around 300 pounds. *Mechanical Engineering* magazine editor Alan Brown in a 2007 article opined, "You could walk into any gym and find a dozen men and women who could bench press one of them."

The PW600 would be built in a series of modules so that each could be tested, maintained, and switched in and out individually. The rugged little powerplant would have about 50 percent fewer parts than the PW500, again focusing on cost and maintainability. At first it appeared Pratt & Whitney Canada was a bit behind in this emerging market. Williams really had pioneered very light jet engines, spinning off its cruise missile engine development. But the Williams engine did not meet the requirements of the first VLJs. Because Pratt & Whitney Canada had run a very successful technology demonstrator program, there was minimal risk as it stepped into the market even as Longueuil was still developing a production model. Wins would come on the Cessna Mustang, the Eclipse 500, and the Embraer Phenom 100. This gave Pratt & Whitney Canada an extremely solid market position. One of the more intriguing competitors emerging as of this writing is HondaJet, developing an engine in conjunction with General Electric.

Well, Pratt & Whitney Canada had won the business, but how was it going to build all of those small, relatively inexpensive engines. By this time United Technologies had adopted a methodology called ACE: Achieving Competitive Excellence. It is derived from all of the best in lean manufacturing principles exemplified by Japanese engineering and manufacturing, including the Toyota method. For the PW600 Pratt & Whitney Canada used ACE principles extensively, especially what it calls 3P, production preparation process.

In an interview with *Aviation Maintenance* magazine, Vice President of Customer Support Maria Della Posta described this new way of thinking as the "to be" process. Think about and plan for what would be the best experience for the customer. "We want to shed the traditional view of how it *has* to be," she said, "kind of thinking in the customer environment."

As production ramped up, the goal was to assemble and test an engine in eight hours. By way of comparison, the big engines south of the Canadian border could take several days. Pratt & Whitney Canada and the VLJ builders are blazing a trail in how the industry thinks about building things. Certainly, Pratt Canada's innovative and creative approach to a whole new market is helping all of Pratt get ready for the next 80 or so years.

# Great Engines
## and the Great Engine War

## chapter 8

**PRATT & WHITNEY** had great success with its J57/J75 engines in powering the fighter aircraft of the 1950s, but its main rival, GE, had also come up with a winner in the J79. It powered the F-4 Phantom that had become a mainstay of the Air Force, Navy, and Marine Corps and was garnering strong international sales.

An opportunity for a new engine for the Navy, Pratt's traditional customer,[1] emerged in the late 1950s called the Missileer. It was to be a high-altitude carrier air defense plane. The idea was the Missileer would orbit high out from a carrier battle group, armed with advanced, long-range missiles to engage approaching enemy aircraft.

Pratt & Whitney won the engine competition with its JTF10 that in military parlance became the TF30. It was the first turbofan, afterburning engine the company had ever tackled. Both fan and engine core exhaust were mixed in the afterburner. This set the pattern for future fighter engines. You could build an engine with a relatively small core that reduced weight and drag and low bypass ratio (1.1 to 1 in the TF30) and still produce a lot of thrust in the afterburner. The turbofan improved fuel burn some 25 percent over the traditional turbojet design of the J57/J75. The afterburner was throttleable, which allowed a

The J-52 would be a work horse military engine for decades.

[1] For many years Pratt was labeled a "Navy installation." Navy procurement officers supervised all military work no matter who the ultimate customer. The practice of individual services supervising a plant ended with the creation of the Defense Contract Administration Service (DCAS), whose multiservice personnel were sometimes referred to as "purple suits."

pilot to select thrust increases up to 70 percent over the nominal rating of 11,350 pounds to 13,400 pounds. In afterburner you could go as high as 25,100 pounds in later TF30 models.

The Missileer, of course, never happened. By 1961 newly sworn-in Secretary of Defense Robert McNamara wanted the Air Force and the Navy to join in a single program, the Tactical Fighter Experimental or TFX. Neither service nor any of the potential contractors were enthusiastic about the idea, believing that you couldn't make a "one-size-fits-all" aircraft that could be a land- and carrier-based fighter and an interdiction bomber with nuclear capability. But the TFX was the next major program. McNamara was determined that the Pentagon would use it as a template for wide-ranging reforms in procurement. Two models emerged when contacts were let in 1962, the F-111 from General Dynamics for the Air Force and the F-111B from Grumman for the Navy. Pratt & Whitney won the engine for both with the TF30. The teams were given 25 months to get the aircraft into production.

The schedule was extremely tight and the requirements very difficult. The Air Force needed a relatively heavy, robust airframe because the F-111's interdiction role called for nap-of-the-Earth capability. The Navy, as always, wanted a much lighter aircraft for carrier duty. And the F-111B was getting heavier as was the Air Force F-111. This meant real challenges for the people working on the TF30. Issues in performance revolved around weight and the somewhat contradictory requirements between land-based and carrier-based aircraft.

The whole TFX/F-111 program became extremely controversial and contentious on Capitol Hill, in the Pentagon, and in the offices of the contractors like Pratt & Whitney. It is not the intention here to go into detailed account of the whole saga. Suffice it to say that Pratt & Whitney had to scramble to get the engine right, and it did after constant updating. But by then there was no F-111B for the Navy. The carrier aviators dropped out of the program because they did not believe they could get a suitable aircraft. The Navy would go its own way with its VFX program at Grumman. The F-111/TF30 would go on to a long career with the U.S. Air Force and is only now being withdrawn from service by the Royal Australian Air Force. The engine, in a nonafterburning version, was also used on the Navy's A-7 and was in production for 22 years until 1986.

# DOGFIGHTING MAKES A COMEBACK

As the 1960s turned into the 1970s, U.S. military aviation was going through a reassessment of where it needed to be. During the Vietnam War, it was becoming clear that Air Force and Navy tactics had let dogfighting skills slip. Over North Vietnam the U.S. was losing one airplane for every 2.5 enemy losses. In Korea it had been 12 enemy losses for every American. Critics pointed out that the mainstay F-4 did not even have an air-to-air gun originally. It and planes like the F-105 Thunderchief were superb, but not what might be needed down the road. New Soviet aircraft such as the MiG-23 Flogger and MiG-25 Foxbat would be produced in great numbers and looked formidable indeed. The Air Force and the Navy both were looking for clear winners in the air supremacy battle. The Air Force FX program led to the F-15, and the Navy's VFX led to the F-14, after abandonment of the naval F-111B. In 1970, after a fierce competition with GE, Pratt & Whitney won the contracts to power this new generation of very high-performance combat aircraft. The Air Force designated its engine the F100, and in an effort to save funds the Navy agreed to use a derivative dubbed the F401.

A TF30 lights up the Florida evening.

93

This program was going to be very tough sledding any way one looked at it. The military wanted an engine that was nearly at or perhaps over the edge of anything that had gone before. The thrust-to-weight ratio would have to be at the unheard of level of 8-to-1 when the best up to that point had been 4-to-1. The 25,000-pound thrust class F100 would weigh in at about 3000 pounds compared with the 17,300-pound thrust J79 that tipped the scales at just over 3800 pounds. Pratt & Whitney started looking into new materials to accomplish all of this, much more titanium and high-nickel content alloys than had ever been used before. Directionally solidified blades would be needed in the turbine. The development program did have its difficulties. Components failed, were redesigned, fixed, and tried again. The Pentagon was concerned because its reputation and its budget were slipping as the Vietnam War wound down to its sad conclusion. And the schedule for the new F-15 was very tight. The Air Force needed the plane to meet the threat of the already deployed Floggers and Foxbats.

66 **The F-15 and F-16 are, beyond any doubt, two of the highest performing fighters in the world . . .** 99

The Navy killed the F401 after some initial test failures amid an ever-constricting budget for naval aviation. The Navy now had an aircraft on the way, the F-14, and no engine in a very tight budget. The option was to go with the best available powerplant even if it wasn't perfect. The Navy picked the Pratt & Whitney TF30, although no one, even Pratt engineers, believed it was the ideal solution even if next to the F100 it was the most advanced fighter engine in the world at the time.

But the Pratt & Whitney team kept working as furiously as they could on getting the F100 into service on the twin-engine Air Force F-15, and they did. Also, in 1974 the engine won the competition for the new single-engine F-16 fighter. But shortly after the F-15 went into service, difficulties arose. This was partly because of the greatly expanded flight envelope the F-15 and the power of the F100 opened up for pilots who loved their new warbird. By 1974 the aircraft held eight world records and astonished everyone in aviation with its ability to accelerate right past Mach 1 while flying straight up.

In 1979 congressional testimony, General Alton Slay, head of the Air Force Systems Command, gave the aircraft and engine high praise:

> The F-15 and F-16 are, beyond any doubt, two of the highest performing fighters in the world, if not *the* [emphasis in original text] two highest performing fighters in the world. This very salutary fact is due in no small measure to the impressive performance of the F100 engine. . . . Without doubt, the total engine design pushed the state of the art.

There was, however a "but" in that testimony by Slay: "The engine has met our requirements in each of three key performance parameters we set for it: thrust, weight and specific fuel consumption; however, it has not yet met our needs for durability nor for ease of maintenance."

When the F-15 was first envisaged, it was thought that most flight time would be at fairly steady throttle settings for long periods of time. But as more and more data from Vietnam and Soviet aircraft studies became available, the mission profile changed. And as F-15 pilots started getting used to their new, incredibly capable aircraft, they found it could do things they had never envisioned possible. There would be a lot of throttle movement, especially "snap acels" from flight idle to full military power in afterburner and back again.

In many cases the violent moves that pilots could now do that had not been foreseen in the original aircraft/engine design led to compressor stall/stagnation. The airflow through the engine was interrupted. Thrust fell off sharply while temperatures, already high in this advanced engine, shot up even more. This in turn caused a lot of premature wear and sometimes failure in the turbine. In some cases, engines had to be shut down and relit.

Traditionally, when engineers designed a new powerplant they looked at a number of engine starts, a number of takeoffs, a number of afterburner lights, and a percentage of time at high power as the key parameters. In the dry language of a 1980 Society of Automotive Engineers technical paper, Pratt summed up the issue:

> It was later realized that one of the most important duty cycle parameters—the number of power lever excursions from low-to-high-to-low power—was not specified. These power lever transient excursions have a major impact on engine life. The engine industry did not fully recognize at that time the impact of this parameter on engine design life. The F100 engine experiences loads and cycles which are significantly different from those for which it was designed.

The Air Force's original estimate, according to General Slay's testimony, was that a typical F100 engine would see 1.5 thermal cycles per flight hour. It turned out the F-15 was seeing an average of 2.2 and the F-16 3.1. Again, in the matter-of-fact language of that 1980 technical paper, "The resulting increase in the rate of thermal cycle accumulations increased thermal fatigue with a corresponding reduction in turbine module durability." That is to say, turbines wore out faster than they were supposed to. Pratt & Whitney and its suppliers could not keep up initially. The F100 was getting a bad reputation for "stall stagnation."

It was one of the toughest periods in the history of Pratt–U.S. Air Force relations. Communications between the two became extremely strained in a web of contractual issues. A component improvement program eventually got the problems solved, but Congress was unhappy. The post-Vietnam angst at the military was still very strong, and there were all kinds of sensational news stories about seemingly wildly expensive defense procurement. The Air Force was getting a lot of heat as was the whole military on the life-cycle cost. And it could be said fairly

The F100 represented a breakthrough in engine technology.

that Pratt & Whitney got into a kind of bunker mentality. It just was not listening to which way things were going. And thus began "The Great Engine War" between Pratt & Whitney and General Electric over which company would power future F-16s.

Certainly GE had been watching all of these developments, and it had never given up after losing the first round in the F-15/F-16 engine competition. It had the engines for the B-1 bomber and was on its way to get the F-18 for the Navy. It showed that its F101 Derivative Fighter Engine worked well in the F-15 and F-16. The Great Engine War to power future F-16s generated news stories, magazine articles, and congressional testimony. Claims and counterclaims filled volumes. But in the end it was a devastating loss for Pratt & Whitney. In February 1984 the Pentagon announced that the GE F110 as it was now called would get 120 F-16s in the next buy and Pratt & Whitney would get 40. Adding insult to injury, the Navy decided in 1987 to re-engine its F-14s with a GE F110 derivative, never having been satisfied with the TF30.

Dick Coar, the plain-talking Pratt & Whitney president of the mid-1980s, summed up the bitter lesson of the Great Engine War succinctly: "We needed to pay more attention to the Air Force. All our money comes from customers. We don't print it and if we don't satisfy customers we won't get it."

The late Brian Rowe, head of GE Aircraft Engines, might have been Pratt's toughest competitor over the years. His 2005 autobiography, *The Power to Fly*, had his own blunt analysis: "We had some help from Pratt & Whitney who were more interested in stopping us than in fixing their problems."

A 1984 postmortem article in *Dun's Business Month* quotes an anonymous Air Force general's tough summary of the Great Engine War: "History impresses people more than philosophy. I'll tell you one thing. The next time the Air Force tells a contractor to pay attention to something, they will. That's what this did. It got their attention."

The people of the Government Products Division, stunned as they were, certainly were paying attention. Bill Missimer, who by 1990 would rise to run the division, said in a message to employees after the loss: "We're going to set new standards of excellence with the Dash 220 (the F100-220, an improved model). We're going to get it into production without a hitch and produce it with outstanding quality on schedule." And the company did, winning back more and more of the F-16 market for both the U.S. and international customers. Customer support and just plain listening became the top priorities. The F100 became the most reliable, safest engine in the U.S. Air Force (USAF) inventory. There are more than 7000 F100s in service with the USAF and 22 other air forces accumulating well over 21 million flight hours.

Missimer looked at the lessons learned in the Great Engine War:

> I think we have changed from a company with a personality that was not responsive to a company that is responsive. It's not one change. It is many things that make a difference. How we react to the customer is key. We also had this foxhole mentality that when we got in a difficult situation we didn't want to come out and talk about it until we had all the answers. That was the wrong approach. Now we know we have to keep our customers fully informed.

The F100 program, despite the disappointment of the Great Engine War, taught Pratt & Whitney valuable lessons and moved its technology far ahead, especially in areas of aerodynamics, materials, cooling, and coatings. It did another thing. It helped the company learn how to live and work internationally.

The NATO fighter program of the 1970s, sometimes referred to as the Arms Deal of the Century, had F-16s in Norway, Denmark, Belgium, and the Netherlands. But to win General Dynamics, the F-16 contractor and Pratt & Whitney had to agree that 40 percent of the value of the 300-plus airplanes would be built in the partner countries. Pratt & Whitney people from West Palm Beach and East Hartford all of sudden found themselves living in Brussels and traveling all over Northern Europe to find partner companies like Fabrique Nationale, Norsk Jetmotor, Eldim, and others. Since 1925, Pratt & Whitney had pretty well designed and built all of its own products, albeit with a big supplier base. Now it had true partners in the engine business outside the United States. It was a pattern that would grow and develop.

The first F-15 takes flight with two F100s.

Photo courtesy of The Boeing Company.

Today the company has a network of international risk-sharing partners and preferred suppliers all over the world. You can view the NATO F-16 program as the first step toward Pratt & Whitney's becoming a truly international company.

If those who had struggled through the Great Engine War needed vindication, it came in 1991. The Air Force was picking an engine for the Advanced Tactical Fighter (ATF) that would become the F-22. There were those who figured the engine war had cooked Pratt's goose or, waggishly, its eagle. They were wrong. The Pratt & Whitney F119 won, and within a couple of more years its derivative, the F135, would be chosen as the lead engine for the Joint Strike Fighter, the F-35, the biggest military program in aviation history in terms of value.

With apologies to Winston Churchill, Pratt & Whitney folks could perhaps appropriate his comment about France's feeling that England would not last long after Dunkirk. The world predicted that GE would wring the Eagle's neck like a chicken. Some chicken! Some neck!

F100-powered F-16s serve around the world with more than 22 air forces.

Photo courtesy of Lockheed Martin.

**98**

# New Engines,
## New Challenges

## chapter 9

**AS THE 1970S WORE ON** and moved toward the 1980s, the commercial engine world that Pratt & Whitney dominated with the JT3, JT4, JT8, and JT9 was changing in fundamental ways.

Since the beginning of jet aviation, the general rule had been that each airplane had only one type of engine installed, and Pratt had dominated that market. The widebody era changed that with GE and Rolls winning positions on the DC-10, L-1011, and 747 and eventually finding a home on the new Airbus A300 and the Toulouse-built aircraft that followed. This in turn led to severe price competition among the three major engine builders, which continues to this day. The era of fleet introductory assistance, steep discounting, had arrived. Each of the three at one time or another has blamed the other two for starting the intense price competition, and it is probably a "chicken and egg" argument that has no clear answer. Suffice it to say, the rules of the game changed. In most cases engine companies lost money on initial sales and made it up over the years through the sale of spare parts. Jet engines became "the world's largest razor blade business" as the catch phrase goes.

In a 1987 *Aviation Week* interview Larry Clarkson, then president

The JT10D, the one and only.

of Pratt & Whitney commercial engine operation, summed up the situation that had developed: "Using 20/20 hindsight, not offering a launch engine for the DC-10, L-1011 and A300 was a mistake," he said. "It allowed our competitors to get major footholds in airlines that had traditionally been Pratt customers. It also pressured Boeing to eventually offer our competitors' engines on the 747."

Another change that was coming, as we noted earlier in discussing the European fighter engine program, was companies outside the United States were beginning to win larger roles in engine programs. For Pratt this started with the F100 NATO program and then went on in the development of the JT8D-200 where MTU of Germany and Volvo of Sweden became risk-sharing partners. They invested in engine development and then shared in the profits when they came, sometimes years after entry into service. At first these partnerships were relatively small with one of the big three teaming with smaller companies. As time went on and the engine development economics became ever tougher, each of the three would form partnerships with one another. Individual companies could no longer afford to go it alone on a program that might run up a billion or more dollars in development costs before the first sale. Profits would not appear for years until significant spare parts sales appeared. Better and much more durable engines meant the old rule of thumb that an engine company sold the engine three times at least—once initially and twice more in spares—was not necessarily true any longer.

There would be CFMI, a 50/50 joint venture between GE and SNECMA of France. Then came International Aero Engines (IAE) with Pratt and Rolls each having about a third interest with MTU, Fiat, and Japanese Aero Engines splitting the remaining third. GE and Pratt & Whitney would later form the Engine Alliance as a 50/50 partnership to develop an engine for the Airbus A380. Rolls and GE would form an alliance to develop an alternate engine for the Joint Strike Fighter. And even engines that carried one of the big three's nameplates had significant partnership shares. All three companies would find some difficulties in making these new business and technical partnerships work smoothly, but they did, and this has become the pattern of the commercial engine business for the 21st century.

# THE JT10D – THE ONE AND ONLY

Pratt & Whitney's first attempt to form a truly international consortium was the JT10D program of the mid-1970s. Boeing was looking at new aircraft for the 1980s that would replace the 727, the best-selling jet aircraft in history at the time, and possibly the 737, which, believe it or not, was something of a stepchild in those days. Pratt came up with a 25,000- to 35,000-pound class of engine dubbed the JT10D, which would enter service rated at 26,000 pounds of thrust.

Pratt would own 83 percent of the program with MTU of Germany taking 13 percent and Fiat of Italy 4 percent. What really got the aviation world talking was the decision in 1976 for Rolls-Royce to join the program and take over 34 percent. Yes, Pratt had worked with Rolls back in the 1940s with the Nene program, but that was strictly military. For the first time the two rivals would be working together in a significant commercial venture.

But by May 1977 the engagement was broken in what Pratt described as "a natural and amicable decision" for each company to go their own way. The chief reason was that Boeing's planned airplane kept getting bigger and requiring more thrust. The initial JT10D was now going to be a 29,000-pound thrust engine, not 25,000 or 26,000. That thrust requirement was getting very close to what Rolls was doing with its RB211 for the widebody aircraft. Rolls clipped the RB211 fan and came up with the RB211–535 at 32,000 pounds of thrust initially.

"When Rolls decided to build a competitive engine they were indicating that they did not want to participate in the JT10D as it stood," Pratt & Whitney President Bruce Torrell told *Aviation Week* magazine. "Instead of a collaborator, they became a competitor. There is no animosity toward Rolls. We're just following a different path. We may at some time in the future collaborate again."

FADEC's, electronic engine controls, would change the whole industry.

And despite some bruised feelings in both companies, they did so with the creation of International Aero Engines to develop the 25,000-pound thrust class V2500. But here is where things get a bit complicated.

Pratt and Rolls needed the V2500 because they had not developed the original JT10D.[1] Boeing wanted more thrust for its new plane, which would become the 757. Pratt's JT10D would morph into the PW2037, and Rolls would push the RB211–535. GE dropped out. The 757 was the plane to be on. After all, it would be the next 727, a big seller. With airline deregulation in the United States and the development of hub and spoke airline systems, however, the stepchild 737

---

[1] There is one remaining JT10D. It is in the Pratt & Whitney museum in East Hartford.

took off and became the hottest seller in commercial aviation, which it is to this day. Boeing wanted to improve the airplane and needed a new engine with more thrust than the Pratt JT8D, which equipped the original 737s. Pratt and Rolls were committed to the 757 and just did not have the resources to go after the new version of the 737. Although in hindsight the success of the 737 seems self-evident, it was not at the time when the effects of deregulation were not crystal clear.

In any case, Boeing opted to build the 737–300 with an engine that had only been used to reengine old DC-8s. It was called the CFM56, and it was built by CFMI, that joint venture between GE and SNECMA of France. SNECMA had flirted with Pratt & Whitney initially, and, in fact, Pratt took a 10 percent ownership in SNECMA at one time. But no real engine collaboration ever emerged. At the time SNECMA did not want to cozy up to Rolls Royce, mostly because Rolls was in deep trouble with its engine for the L-1011 and was in essence broke. That left GE. The two got together in 1971 and developed the CFM56. They did not sell one for almost eight years, and the first sales were to reengine DC-8s. Then Boeing came up with a plan to use the powerplant to reengine KC-135 tankers. To extend the life of the 737, the -300 model was designed, and the CFM56 would be its engine. And that in turn would give CFMI a big leg up when Airbus unveiled its single-aisle A320, the airplane which Pratt and Rolls were aiming at with the V2500.

So the decision in the mid-1970s to go for higher thrust, made for all of the right reasons, has had a profound effect on Pratt & Whitney and the whole engine industry. In the many years since then, the CFM56 has accounted for more than half of all large commercial engine sales and passed the JT8D as the most successful commercial jet engine in history.

# THE 2037 – THE ENGINE THAT THINKS FOR ITSELF

As the Boeing 7X7 moved to become the 757, not only was increased thrust an issue, but far more important was fuel burn. Originally jet engines were attractive because they burned a cheap kerosene derivative, much less expensive than high-octane aviation gasoline for piston engines. The first oil shock of the early 1970s changed all that. The cost of jet fuel went up as much as 10 times and was becoming up to half the operating costs of airliners.

Pratt was determined that as the JT10D became the PW2037 it would be the most fuel-efficient ever built. The engine would be 30 percent better than older turbofans and 7 percent better than competing engines like the RB211–535. It would take breakthrough technology to do it. Much of that had been in the testing stage in the 1970s and had been introduced on a limited basis in the final JT9D-7 series engines.

The PW2000 was the first of a new generation of engines from Pratt & Whitney. Its military version, the F117, powers the C-17 airlifter.

One of the biggest changes would be full-authority digital electronic engine controls (FADEC). Today every engine by every manufacturer has them, but when Pratt, working with sister UTC division Hamilton Standard, introduced them on the 2037, it was a radical change. Up until then, the fuel flow to an engine was handled by a hydromechanical device of gears, cams, springs, and levers that connected to the throttles in the cockpit. As engines got more powerful, the controls got more complicated. The hydromechanical control for the original JT9D was a huge device almost as complicated as the engine itself, and it did not offer the precise control engines would need for better and better performance. In fact, some of the issues with the original JT9D could be attributed to the fact that the engine control just could not make the fine, split-second, adjustments the engine needed.

The FADEC was essentially a briefcase-sized digital computer strapped onto the engine capable of calculations and adjustments in millionths of a second. The control allowed pilots to set and forget the throttle with the engine automatically adjusting to changing conditions. And the control could be tailored to a particular airline's operating conditions. The FADEC also could store data for later read-out and troubleshooting before a condition deteriorated into a major maintenance issue. FADEC was a harbinger of the onboard, in-flight diagnostic systems that exist today and are directly linked to a ground station in real time.

The FADEC also helped another major innovation, active clearance control. The computer could decide in milliseconds just how much cooling air would be sent into a series of tubes around the engine's turbine case. As the case heated up and expanded or cooled and contracted, the active clearance control feature meant that the space between the turbine blades and the case could be controlled within the thickness of a human hair or two. This kept leakage of the gas stream around turbine stages to a minimum, greatly improving efficiency. And the turbine blades incorporated abradable tips, so that they, in effect, wore their way into the engine case, again limiting leakage.

The advent of computer-aided design helped create controlled diffusion airfoils. Air moved much more efficiently with separation of the air from the airfoil greatly reduced. And airfoils were thicker and more durable. Single-crystal turbine blades were much more durable because they were in essence one metal crystal. There were no microscopic boundary lines between the myriad metal crystals of a traditional design. Disks were made stronger by using powdered metal. Stronger disks meant that you could spin them as much as 30 percent faster than in previous engines, thus getting more work out of each stage in the engine. That meant fewer stages and fewer parts for a given amount of thrust when compared with earlier engines.

For an engine incorporating as much new technology as the 2037, the development program went smoothly, and the engine flew on the 757 for first time in March 1984. Anytime an engine flies for the first time, it is a special day for the Pratt men and women who have put their heart and soul into it. In a company newspaper article at the time, 2037 project engineer Ed Donaghue seemed to be speaking for the legion of "aircrafters" who had witnessed those first flights:

> Watching the 757 roll down the runway and take off was an exciting moment," he said. "We've been with this engine from the very start. We've seen it through design and assembly, through all the ground tests, and knew it could do the job. But seeing it on the wing and flying felt really good.

The 2037 would face stiff competition from the Roll Royce -535 but held its own even as the urgency over fuel burn died down somewhat after entry into service. As back in the days of the Wasp, the engine was constantly improved to make it even more robust and durable.

In one sense the 2037 proved its value not in commercial service but in places like Afghanistan, Kosovo, Kuwait, and Iraq. In 1983 the U.S. Air Force selected a 40,000-pound thrust version of the engine, the F117, for its new C-17 airlifter. Since entering service in 1991, the airplane/engine combination has proved itself time and again in some very tough conditions. It has carried troops, tanks, and trucks into tight, unimproved strips in remote areas. The engine has performed, to use Rentschler's phrase once again, "like a thoroughbred."

The F117 program was also accomplished under some unique conditions. It was essentially a commercial deal. The Air Force bought an off-the-shelf engine with only minor modifications on commercial terms. Even the engine servicing is done that way by Pratt & Whitney and United Airlines.[2]

The 2037 was also responsible for one of the stranger odysseys in Pratt history: working with the Russian aircraft industry. Since the late 1940s, if a Pratt person even talked with a Russian, he or she had to immediately report it to internal security. At international air shows there were always stories, never authenticated, of thieves in the pay of the KGB trying to steal briefcases of sensitive data.[3]

But by the late 1980s *glasnost* and *perestroika* were opening some doors, even if they were only slightly ajar. Pratt executive Bob Rosati, who headed International Aero Engines twice, made contact with the Soviet industry, especially the Ilyushin design bureau. Out of all that grew the IL96M program. Ilyushin actually sawed an IL96 in two, stretched it, and outfitted it with a Honeywell cockpit, Collins avionics, and four PW2037s. It was a big story at the Paris Air Show and other international air shows in the early 1990s, and some airlines expressed an interest in the plane as a freighter. Certification issues and the chaotic state of the Russian economy and aviation industry meant the program never really took off, nor did a similar effort with Tupolev for a twin-engine Tu254.

105

---

[2] Some interesting sidelights to the 2037 story are recalled here: little remembered now, it was proposed for a twin-engine version of remanufactured 727 trijets; also it has been put up a few times as a four-engine replacement for the eight engines on the B-52.

[3] It was tradition at Pratt that the company would fly the national flag of visiting foreign delegations. The head of security was on hand the day the hammer and sickle first went up, muttering, "I never thought I would live to see the day."

The 94-inch fan PW4000 made significant fuel burn improvements over earlier generation high-bypass engines.

# THE NEXT BIG THING

As development work continued on the 2037, Pratt & Whitney was looking at all that technology and what they should do with it next. The JT9D had ushered in the high-bypass turbofan era and had done well in the marketplace. The latest versions of the engines, the 7R4s, were quite popular in the early 1980s. But GE's CF6 series and the RB211 from Rolls were gaining. The JT9D was up to 56,000 pounds of thrust, and growing it further was problematic. It was an expensive engine to build, and that was a consideration as the competition for widebody business reached white heat levels.

With much fanfare on December 8, 1982, UTC Chairman Harry Gray along with Pratt & Whitney executives announced that the company would develop the PW4000, a brand new high-thrust engine for widebody aircraft.

Bill Robertson, head of JT9D and PW4000 programs, told *Aviation Week* the thinking behind the decision: "The JT9D could have been made to reach 60,000 pounds of thrust, but we needed increased engine life, reliability and lower operating costs. With all the changes needed to accomplish this, we decided we might as well make a new engine."

Pratt commercial engine business head Larry Clarkson told that December 8 audience that the company had been investing in new technology for 10 years, much of it at that moment on test with the 2037. Now it was time to move it to widebody aircraft.

"The greatest benefit from all of these innovations comes only when you give the engine designer a clean sheet of paper and tell him or her to make a fresh start," Clarkson said.

All of the technology just mentioned on the 2037 and then some were incorporated into the design of the PW4000. It would be a third-generation commercial jet engine. For instance, the engine rotation speed was increased to 10,000 rpm, 2000 faster than the previous JT9D. This allowed for more work to be wrung out of fewer engine stages. The result was a 30 percent reduction in high-pressure compressor airfoils and a 50 percent drop in high-pressure turbine airfoils.

This, combined with improved fuel burn, lowered the cost of the engine and the cost for airlines to operate and maintain it. And cost was a big issue. It was a theme Pratt people were taking to heart as well as their colleagues at GE and Rolls. Getting cost out of engines was and is a daily struggle for engine builders in the age of intense competition and Wall Street's demand for outstanding financial performance.

For the PW4000, engineering and manufacturing were brought together in a new way so that, in Pratt speak, designs would not "be tossed over the transom to the shop."

PW4000 leader Bill Robertson explained the drive toward becoming a low-cost producer: "Engineering was given the task of designing to cost and it joined forces with people in the Manufacturing Division who are experts in determining the most cost-effective way of building an engine."

Jim Bruner, the engineering manager for the PW4000, explained that manufacturing and engineering people were brought together before the first line was drawn on the first piece of paper.

"There was a lot of brainstorming in those early days," he said. "We left no rock unturned and questioned every aspect of our past approach to design and development. We collectively established a program motto, 'business as usual won't work.'"

The first order for the PW4000 came in late 1984 when Pan Am picked the engine for its new fleet of Airbus A310s. In hindsight, it was a bittersweet moment. Pan Am had begun transoceanic service with Pratt & Whitney –powered

PanAm's PW400–powered Clipper Pratt & Whitney.

Sikorsky clippers and then Boeing Stratocruisers with Wasp Majors. It launched the JT3 and 707 and the JT9D and 747. This would be the last PanAm launch order. By the early 1990s Pratt & Whitney tractor trailers were dispatched to Kennedy airport to pick up engines and parts that PanAm, headed for bankruptcy, could no longer afford to own.

The PW4000 went into service at 56,000 pounds of thrust and within a few years was certified up to 62,000 pounds for some applications. It would find homes on the Airbus A300/A310 family, the Boeing 747 and 767, and the McDonnell Douglas MD-11. The engine always has faced intense competition from the excellent GE CF6–80C2 and Rolls Royce RB211. The PW4000 was also selected to power the Boeing 767 entry into the U.S. Air Force's aerial tanker competition.

When the PW4000 went into service in 1987 at the Paris Air Show, everybody thought a 96-inch fan engine that could go up over 60,000 pounds of thrust was about a big as you could get. Like so many times before in aviation, we all forgot the sky really is the limit, and there were even bigger things to come.

# THE JT10 DIVORCE WINDS UP IN A SECOND MARRIAGE

Although Pratt & Whitney and Rolls-Royce had parted ways on the JT10 project, the need for a new 25,000-pound thrust engine was still there as the 1970s moved into the 1980s. The 737 had taken off, and Airbus was developing its competitor, the A320.

Pratt had begun looking for a way to develop an engine below the PW2037 as early as 1979. In a 1985 *Exxon World* magazine interview Bob Rosati, the man who would head International Aero Engines (IAE), explained the thinking: "There are lots of types of commercial aircraft, but there are only three types of engines. Like sweaters, they come in large, medium and small." Pratt had its large engine in the PW4000 and its medium engine in the PW2000 series. It needed a small one, but financial and technical resources were stretched to the limit. "When we got to the small engine," Rosati recalled, "we found we needed a huge investment and that the risk was not commensurate with the return. Yet we did want a full product line. So the way to do that was to go into partnership."

Rolls-Royce was facing similar resource issues. In the same interview Sir Ralph Robins, head of Rolls-Royce, described how his company had been working with the Japanese on a 25,000-pound class engine, the RB432. "We got the Japanese into that and were working with them about four years before we decided we needed a wider collaboration."

Who first proposed this new marriage? "To be absolutely precise as to who spoke to whom and when on this project—I just don't remember. It just happened," Robins said. Bob Rosati recalled that it took a couple of years of study and negotiation to get things going.

In March 1983 seven companies signed the collaboration agreement that created International Aero Engines and the V2500 series. Rolls and Pratt each held a 30 percent share with MTU of Germany, Fiat of Italy (which has since withdrawn as a partner but remained a supplier), and Japanese Aero Engines Corporation (JAEC) dividing the rest. JAEC itself was a consortium of Mitsubishi, Kawasaki, and Ishikawajima-Harima.

> " We took 27 engineers from various companies and locked them in a building and told them to settle it. "

Pratt & Whitney would develop the diffuser, the combustor, and the high-pressure turbine. Rolls would have the high-pressure compressor. MTU got the low-pressure turbine. JAEC would build the fan and low-pressure compressor while Fiat did the turbine exhaust case and gearbox. The new outfit would be almost a "virtual" company with only a 170 employees, all on assignment from partner companies. IAE headquarters would be a converted schoolhouse next door to Pratt headquarters in Connecticut. No one had ever tried a collaboration with this many partners spread out all over the world. Bob Rosati called it "a milestone in international collaboration." Many others were less kind. How would these disparate companies, cultures, and countries work together?

Out of the first 100 interviews only 16 people were picked. Rosati wanted people who would become IAE not Rolls or Pratt or MTU or Fiat or JAEC. Like so many times before in aviation, people's passion overcame their name badge, and they did, indeed, become IAE people.

"For example, we had a big argument at the beginning on what type of engine this would be," Rosati recalled in that 1985 article:

> We took 27 engineers from various companies and locked them in a building and told them to settle it. Within three days you wouldn't have known who was Rolls or who was Pratt, except for the accents. They got wrapped up in the job. The Rolls guys came out and said, "These Pratt people really know what they are talking about." And the Pratt guys came out saying, "These Rolls guys are pretty smart after all."

The V2500 is a success story of international cooperation.

This is not to say that there were not bumps along the way and some fairly good sized ones. You had to bring parts and components from seven partners and dozens of suppliers to Connecticut and Rolls-Royce in Derby, United Kingdom, for assembly into a working engine. There were problems with the high-pressure compressor that led to a major redesign effort, giving the low-pressure compressor three stages instead of one to ease the workload on the high-pressure's 10 stages. This issue and others led to the partners assigning overall engineering responsibility to Pratt. The structural change was to get greater efficiency and focus rather than have each partner run an independent, full-up engineering organization for the V2500.

The V2500 went into service in May 1989. It has had outstanding success on the A320 family, where it competes head-to-head with the CFM56, which got on the airplane first. It is also the powerplant for the McDonnell Douglas MD-90. More than 5000 V2500s worth more than $30 billion have been sold.

A plan to build a "SuperFan" version of the engine for the Airbus A340 never came to fruition as IAE concentrated on certifying and getting into service the base engine, overcoming its compressor development issues. And despite efforts over the years IAE and its partners never won a position on the 737 mostly because of the daunting financial issues of developing and certifying an engine for the Boeing bestseller. But as we shall see later, the ideas of a geared fan engine and getting on every 150-passenger aircraft have never gone away.

# The Super Bowl of
## Fighter Engines and Big Twins

## chapter 10

**PRATT & WHITNEY WAS NOW FACING** a much different marketplace, both in its military and commercial engine businesses. In the 1950s the company had been able to find lots of applications for the J57/J75 family as the U.S. military went into the Jet Age. There were new aircraft programs coming down the pike regularly. If Pratt missed one, there was another right behind it. Some 14 military aircraft would use J57/J75 power. And the same basic engines worked well in the commercial world on the 707 and DC-8. But by the time the next military engine, the TF30, came along, there would only be three aircraft applications. The F100 would have only two homes, the F-15 and F-16. The U.S. military was now looking at only one high-performance fighter in the early 1980s, the Advanced Tactical Fighter (ATF), which would become the F-22. The merry-go-round was turning about once every 20 years now, and one had to catch the brass ring or wait a long time until the next chance.

In late 1983 Pratt & Whitney began the journey that would take it to perhaps the most critical military programs in its history. The Air Force Aeronautical Systems Division gave the company a $202 million dollar initial development contract for the Joint Advanced Fighter Engine. Of course, General Electric got a similar contract. It was another Great Engine War.

The requirements were tough. The engine would have to be able to cruise supersonically without afterburner (supercruise). It had to have thrust vectoring for extreme maneuverability. Its signature could not compromise the stealthy design of the aircraft. The engine had to do all this and more while having great life-cycle cost and ease of maintenance. It is perhaps ironic that a multimillion dollar, high-performance aircraft is usually taken care of by 20-something mechanics with perhaps only a few years or even months of experience.

The internal company designation for the engine was the PW5000. GE's entry was the GE37. After the bitter blow of the Great Engine War on the F-16, the people of Pratt & Whitney were grimly determined to win the competition that began formally in 1983. "The competition is a head-on, day-to-day battle," said Frank Gillette,

PW5000 engineering manager at the time. "Whatever help was needed, people rallied with a 'we're going to win' attitude. And that's why I'd bet my bottom dollar we're going to win."

Military engine executive Rick Silva recalled years later how the people in West Palm Beach and throughout Pratt felt after the drubbing in the F100 Great Engine War. "Okay, beat us like dogs, but we all vowed we'd never do that again. It was the whole company."

Tony Pizzi, who led development operations in West Palm Beach, told his colleagues in a 1985 company newsletter that building the PW5000 was the most important thing on their plate since the F100 development of the 1960s. "We have to 'measure twice and cut once' and produce parts to cost and on time," he admonished. "We all have to work together to do it. That's the bottom line of the entire operation. It's our competitive edge. We'll beat GE because of it." The lessons learned from how Pratt people developed the PW2000 and PW4000 worked their way into the PW5000. Teamwork became the mantra of everyone involved. That transom between engineering and manufacturing began to disappear.

One of the more interesting aspects of the PW5000 program was the emphasis on supportability in the field being as important as the more glamorous performance issues. Supercruise might be the glitz, but wrench turning is the guts. For instance, Pratt engineers went on Blue Two Visits. ("Blue Two" is a nickname for Air Force people, "blue suiters" with two stripes on their sleeves as opposed to the brass.) More than 60 engineers got to do the actual maintenance work that their

The F119 in thrust vectoring mode.

designs would require. Back at Pratt & Whitney, tools would be laid on design drawings to look into accessibility and what kind of skills a mechanic would need to perform a given task.

Line replaceable units (LRUs) on the outside of the engine could not be on top or buried behind each other so that a mechanic could get at them quickly. He or she had to be able to take them off within about 20 minutes with common hand tools. Each LRU would require only one size wrench to remove it. That meant the engine's whole tool kit would need only 11 wrenches. The fasteners that held things together were "captive." They did not come all of the way off the engine to get lost or be dropped into an old coffee can that might tipped over at 3:00 a.m. on a dark, rainy Air Force base line ramp.

Step-by-step Pratt worked the program from design to component rig tests to development experimental engines, ground tests, and flight tests. What emerged was a 35,000-pound thrust-class turbofan with counter-rotating compressors and turbines. There is a three-stage shroudless fan. The high-pressure compressor has integrally bladed rotors. That is, the disk and compressor blades are machined from one forging. This allows for a stronger design with higher rotation speeds as opposed to the traditional compressor disk with its blades attached individually. The thrust vectoring nozzle can move 20 degrees up or down from the centerline and complete a full cycle in a second.

Looking back on the development, Rick Silva summed it up: "We had the world's first stealthy, supercruise, afterburning engine. It was something."

Tom Farmer, who had the systems lead on the engine, said a key to the supercruise capability was learning how to better manage heat loads in the engine with new schemes for the cooling air and advanced coatings to insulate turbine blades:

> It is a turbine blade that is operating well in excess of 2500 degrees where the parent metal melts at 2100 degrees and you want for extended life to be operating at 1400 to 1500 degrees. We just grew the temperature capability of the turbine to let it stay at that very high speed for an extended period of time.

Tension was mounting throughout Pratt & Whitney, especially in West Palm Beach, where the military business was headquartered, as 1990 drew to a close. Everything that Pratt people could do had been done. The last flights of the Northrop and Lockheed ATF prototypes had taken place on December 28, and Pratt & Whitney delivered its final engine proposal on December 31. It was 350,000 pages in 262 boxes and weighed 8000 pounds. It was flown to Wright-Patterson Air Force Base in Dayton, Ohio, on a chartered aircraft. GE had it easier. They only had to use a van to get their proposal behemoth from GE headquarters in Evendale, Ohio, to Wright-Patterson.

The Defense Department set April 23, 1991, as the date when it would announce which companies had won the ATF aircraft and engine competitions.

"We've won the Super Bowl of fighter engine competitions," an exultant Pratt & Whitney President Jim O'Connor told employees on that day. O'Connor had been with former Pratt President Art Wegner, head of all UTC's aerospace businesses, waiting for the call that day from Air Force Secretary Donald Rice. It came at 4:15 p.m. The plan had been that the message would be relayed to John Balaguer, head of the military business in West Palm Beach, who would hold the news until he could announce it at a huge employee rally. It didn't quite work out that way.

66 I heard a loud yell come out of the office and I knew we had won. 99

"I heard a loud yell come out of the office and I knew we had won," recalled Balaguer's secretary, DeeDee Hardert-Payton. "Several people in the hallway broke out in big smiles and started congratulating each other." By the time Balaguer and F119 Vice President Walt Bylciw got to the rally, everyone was already celebrating. There is no official company record, but it must have been a good night for the Hi-Test Lounge.[1]

The Air Force had picked the Lockheed F-22 for the airframe over the Northrop-led team's F-23 and the Pratt & Whitney F119 over the GE120. At the time the Air Force said that the Lockheed/Pratt combination offered the best cost and lowest risk. GE's engine was a very aggressive, variable-cycle design. It would operate as a turbojet at supersonic speeds and as a turbofan at subsonic conditions, significantly improving fuel efficiency.

In its postcompetition assessment *Aviation Week* speculated that the variable-cycle technology might have been seen by the Air Force as too much of a reach: "This added complexity, which may have eventually resulted in greater maintenance requirements, could have hurt GE during the engine selection."

Pratt's Walt Bylciw explained to the magazine that Pratt was confident its more traditional route would do the job. "Pratt has a long history of variable cycle experience with the J58 (SR-71) engine and it was our judgment that there was no need to incorporate this capability—and the weight and complexity it adds—to meet the ATF engine requirements."

Winning the competition was in a sense only the beginning. A lengthy, complex, development and flight-test program would follow. The F-22 Raptor would fly for the first time on September 7, 1997, and was declared operational in December 2005 and mission capable in January 2006. It is the world's only fifth-generation

---

[1] When Pratt first opened in Florida, there was nothing anywhere near it as there is today. The first thing that got going was a little gas station. It became a kind of tradition to stop after work to fill up on gas and grab a beer from the rather large set of coolers inside the tiny place, hence the name the Hi-Test Lounge. In-depth research proved conclusively the lounge sold probably as many gallons of beer as gasoline.

stealth fighter, and its F119 engine is the most advanced gas turbine in service.

But the path to that success was not as straightforward as it might sound today. Tom Farmer recalled that there were plenty of tense episodes. For instance, the engine was under test at the Arnold Engineering Development Center (AEDC) in Tennessee in late 1996 and early 1997. It was getting ready for flight clearance, and the augmentor nozzle was not performing well. A team of engineers went into an office and shut the door. "And they would come out for occasional comfort breaks and we would send in lunch. They stayed in that room and worked that [problem] and finally thought they had it." So a dozen Pratt & Whitney people went up to AEDC to work the fix. "I will never forget going up and seeing that nozzle," Farmer recalled. "Those thousands of parts scattered out across the big concrete floor and these very dedicated people working to implement the design change and get it built back up. They did it. They were spectacular."

When Jim O'Connor called the F119 win the Super Bowl of fighter engine competitions, he could have thrown in the World Series, World Cup, and golf's majors for good measure. It became clear that the engine to power the F-22 would have a leg up to win something called the Joint Strike Fighter. And the JSF would be the biggest military program in terms of dollars in U.S. history, well over $200 billion.

Photo courtesy of Lockheed Martin.

The tremendous power of the F135 and its lift-fan system bring the F35 experimental model in for a landing.

# ONE ENGINE, THREE JOBS

The Joint Strike Fighter (JSF), known formally as the F-35 Lightning II, is the critical program for Pratt & Whitney in the 21st century. The aircraft will replace the F-16, F/A-18, A-10, and AV8 Harrier used by the U.S. Air Force, U.S. Navy, Marine Corps, and the Royal Navy. No doubt many other services around the world will buy it. It represents the potential for thousands of engines. But it is among the most complex programs the company has ever been involved with.

In late 1996 Pratt & Whitney won a competition to develop engines for both the Lockheed Martin and Boeing JSF prototypes. GE had been selected for the McDonnell-Douglas entry, but that airplane did not make the next phase of the competition. The Pratt & Whitney engine would be the JSF119 at this stage, a close relative of the F119 that was in flight test for the F-22. Actually, the engine was four engines. Both Lockheed and Boeing would build conventional-takeoff-and-landing (CTOL) and short-takeoff-and-vertical-landing (STOVL) prototypes. So each company needed two engine models, thus a total of four new engines. And the Pratt teams working on the engines had to have a "firewall" between the group working for Boeing and the group working for Lockheed.[2] The JSF program would also have a huge international component. Rolls-Royce, based on its STOVL experience with the Harrier, would handle that technology. Pratt & Whitney would have to seek other partners and suppliers all over the world to encourage future incorporation of the JSF into those air services.

Whereas the F119 was a 35,000-pound thrust class engine, the JSF engine would be in the 40,000-pound class and would probably need thrust growth beyond that, as seems inevitable in all engine programs. One way of getting the extra power was the tried-and-true Pratt & Whitney method of increasing the efficiency of every engine component possible, getting that last ounce of performance and durability as back in piston days. Also, the engineers advanced the concept they called "supercooling" from the F-22/F119 so that the engine could run at higher temperatures. The idea is to use cooling air taken from the engine's core flow to reduce the temperatures of the high-pressure turbine vanes and blades. Another aspect is using cooling schemes to get very uniform temperatures in the hot gases coming out of the engine's combustor. In a 2001 *Aviation Week* article explaining this system, Bob Cea, head of the JSF program at Pratt & Whitney, said supercooling allowed the engine to run at temperatures about 1000°F above the melting point of the metal used in the turbine blades.

Perhaps the biggest challenge was in developing the STOVL versions of the prototypes. Boeing elected an approach that used direct lift or diversion of the engine's thrust into nozzles that direct it down for short takeoffs and landings. The Lockheed version, which was the eventual winner, used the "liftfan" design. In this engine power is transferred from the engine to a drive shaft and clutch. When the clutch is engaged, the shaft powers up the liftfan that can turn the shaft's 29,000 or so horsepower into about 20,000 pounds of thrust. Air comes into the liftfan through a set of doors behind the cockpit and blows down into a vane box on the underside of the aircraft. That vectors the airstream and can vary the nozzle exit area for lift control. The STOVL system also has roll posts in the

---

[2] In operations there will be a third version of the engine, the carrier variant for the Navy. This engine, however, differs only in detail from the conventional takeoff and landing model.

wings that get about 4000 pounds of thrust from the engine to keep the aircraft balanced in STOVL operation. The STOVL engine also has a swivel module attached to the exhaust nozzle that can vector engine exhaust downward. Getting it all to work flawlessly in seconds was perhaps the biggest challenge of the JSF development. But thanks to digital engine controls, fly-by-wire, and some great engineering (and copious amounts of sweat and blood) by Pratt & Whitney and Rolls-Royce, it worked beautifully and was clearly a superior design over direct lift.

Again, however, the path was not completely smooth. In development tests the clutch for the liftfan system did not hold up well, and the lubrication system for the liftfan gear box was losing fluid. Pratt & Whitney, Rolls-Royce, and Lockheed put together one of those teams that had to sweat bullets and work long days and nights to find a new way of engaging the clutch and fix the lube problem.

But the heart of the JSF engine would always be its cousin, the F119. The high-pressure compressor, combustor, and high-pressure turbine were all taken directly from the F119, which no doubt was one reason the Pratt JSF program went as smoothly as it did. "We will be very careful and judicious in maintaining the benefits of that core engine," vowed Tom Farmer, head of the engine program, in a 2001 *Aviation Week* interview. "We will make very few changes to that section of the engine."

The engines, all four of them, performed well during the fly-off between Boeing and Lockheed. And on October 26, 2001, Pratt & Whitney people gathered in the former hangar at Rentschler Field in East Hartford to watch the announcement of the winner on TV. It was a bit anticlimactic. Whether it was Boeing or Lockheed, Pratt had the engine deal. But now it could be formalized. Lockheed would build the aircraft, and Pratt & Whitney would get $4.8 billion to develop the engine for the F-35 Lightning II Joint Strike Fighter. It was the biggest deal in company history.

The system development and demonstration phase of the effort would have, as all such programs do, ups and downs. The plane would get heavier, the engine would run a bit hotter, the software would not necessarily speak the same language, and technology transfer issues would nettle relationships. Congress would fret about costs and budgets, but the work would continue. It all culminated after more than 10 years when the F-35 took off for the first time on December 16, 2006, and began its flight-test program.

The 94-inch, 100-inch, and 112-inch fans for the PW4000. The engine can produce from 60,000 pounds of thrust up to 98,000 pounds.

The people at Pratt & Whitney had to be proud, but they also had to look over their shoulders. In the wake of the Great Engine War in the 1980s, the Pentagon decided to hedge its JSF bet and fund the GE/Rolls-Royce alternate engine, the F136. As the Pratt & Whitney engine showed what it could do, the JSF program decided it did not want or need an alternate engine and cut the program from its budget proposal. Congress, however, voted otherwise and kept funding it. As of this writing, the issue is still being debated. But what is not an issue is that Pratt & Whitney designed, developed, and built a family of high-performance advanced fighter engines that will be in service for decades.

# THE REALLY BIG SHOW

Long-range and then widebody commercial aviation was built on four- and three-engine airplanes: the 707, DC-8, 747, DC-10, and L-1011. Twins were on short-haul routes in domestic service. But then along came Airbus and the A300, the first widebody twin. It soon had a sister aircraft in the A310. Boeing responded with the 767. Twins started taking over long-haul routes, especially transatlantic service, that had been the province of 747s, L-1011s, and DC-10s. Airlines liked the big twins. Two engines were cheaper to buy, operate, and maintain, and the new two-person cockpits also saved costs. By the late 1980s what airlines wanted was even bigger twins, and they wanted them with extended twin-engine operations (ETOPS) capability.[3] Big twins could then fly the optimum over-water routes, and using a twin instead of a tri or quad made great financial sense. Airlines had been able to fly a twin up to 90 minutes away from the nearest diversion airport allowing for a single engine-out situation. ETOPS certification, however, had come only after many months if not years of service on an engine/airplane combination. It had to demonstrate that its inflight shutdown rate (IFSD) met stringent regulatory standards. What Airbus and Boeing would attempt with their big twins was something brand new: "ETOPS Out of the Box" certification from the first day of service. Engines would be the critical item in making that happen.

The first offering would come from Airbus with a family concept, the twin-engine A330 and the four-engine A340. After IAE's decision not to build a superfan for the A340, the original engine selection went by default to the CFM56.[4] But Pratt, Rolls-Royce, and GE would all bid for the A330 in an increasingly fierce competition in the widebody market.

Originally Pratt & Whitney envisioned the PW4000 as an engine almost identical in size to the JT9D but with capability to comfortably get up to 60,000 pounds of thrust.

---

[3] Wits in the industry with a macabre sense of humor said ETOPS stood for "engines turn or passengers swim."

[4] Rolls-Royce would win a later competition to power longer-range versions of the A340.

The A330 would require more than that, and so the company began the process of upgrading the engine to 68,000 pounds. The low-pressure compressor and the low-pressure turbine would each get an extra stage to supercharge the engine. Fan size was debated, and the team settled on a 100-inch-diameter fan, four inches wider than the fan on the original PW4000. The combustor would be of a new type, TALON (technology for advanced low NOx), in which the air-fuel mixture burned very quickly at the highest possible temperature and then cooled quickly to control nitrogen-oxide emissions. Low emissions and noise had become absolute requirements under U.S. and International Civil Aviation Organization (ICAO) rules. This was getting to be a big engine and weight, which, as usual, was a concern. Pratt & Whitney, working with what was at the time Lockheed Martin's aerostructures business in Maryland, devised an all-composite engine nacelle that for its size was the lightest in commercial aviation. And the PW4168 would be sold as a complete propulsion system, a first for Pratt & Whitney. Previously the company had supplied bare engines with the airframe builder coming up with required accessories and the nacelle around the engine. When the engine went into service in 1994, it was the first commercial engine to have a 90-minute ETOPS rating on day one. It quickly established an outstanding reliability record and won the highest ETOPS available at the time, 180 minutes. GE was competing with a derivative of its CF680C2 family and Rolls-Royce unveiled its new, high-thrust Trent engines.

As the PW4168 was undergoing initial development, another even bigger engine program was getting started. Boeing was responding to the A330 with what was originally called the 767X. Pratt & Whitney discussed offering a 96-inch PW4000, the original engine that could now reach 62,000 pounds of thrust or a version of the 100-inch fan PW4168. Airlines, however, put on a full court press for something more than a stretched and enhanced 767. It became apparent that Boeing would have to offer a totally new airplane, the 777. And Boeing was convinced that Pratt & Whitney's and GE's derivative engines for the A330 would not cut it on the 777. Rolls was offering a new version of the Trent.

There were doubts among Pratt & Whitney engineers that the PW4000 could go up to over 80,000 pounds of thrust. They wanted to use the PW4168 uprated to somewhere in the 70,000 pound or above class. Boeing ruled that out and told Pratt & Whitney leaders that if they wanted to compete they had to come up with something better. The issue for Pratt & Whitney was the torque forces on the engine's shafts at very high thrust. Changing the diameter of the shafts was not a good answer because it would throw off the whole derivative engine design approach. The deadline was fast approaching for a "dribble or shoot" decision, and the engineering department was told to come up with an answer and quickly. They did, using a special alloy and new manufacturing techniques to give the shaft design the strength it needed. Pratt & Whitney was in the 777 hunt with the PW4084, the most powerful aircraft engine ever built at that time for the

biggest twin-engine airliner ever built. The engine would be certified at 84,000 pounds of thrust and eventually grow to 98,000. It would have a 112-inch fan. Another stage would be added to the low-pressure compressor and two to the low-pressure turbine to get the work required to reach those thrust levels.

That big fan posed some unique problems. To build a solid blade out of traditional metal alloys resulted in not only a very big fan, but also a very heavy one. That was especially true if one had the traditional shrouds (snubbers in British parlance) between each blade for strength. All engines fans have to be contained inside the fan case if there should be a catastrophic blade failure from birds or foreign object damage. How could engineers build a case to contain a 112-inch-diameter fan turning at say 9000 rpm and not have the case make the engine very nose heavy? Well, what about Kevlar®, the stuff used to make bulletproof vests? That's what emerged, a thin metal fan case wrapped in many layers of Kevlar fabric.

Now the engineers still had to deal with the problem of the very big fan blades. How would they make them? Engineers remembered that Rolls-Royce had had a terrible time with making big fan blades out of composites for the L-1011. So if solid metal or composites could not be used, what could be used? Manu-

120

The development of hollow titanium fan blades allowed for the 112-inch-diameter fan on the PW4084 for the Boeing 777.

facturing folks, after an incredible effort, came up with a process to build a hollow titanium fan blade that was lightweight yet robust. In essence, they machined two halves and then bonded them together under heat and pressure. The trick was not just in doing it once but in coming up with a process that was repeatable and affordable. The manufacturing team had many long days and nights, but they pulled it off. The big, hollow, shroudless fan blades that the industry was chattering about at the beginning of the program were a nonissue when the first engines went to test.

There was another wrinkle to the PW4084 for the 777. It had to have 180-minute ETOPS "out of the box." The world's biggest engine on the world's biggest twin-engine airliner flying three hours away from an airport over the ocean. This program was going to be watched by a lot of people. And it did not help that just as Boeing and Airbus were moving forward with their big twins the airline industry went into one of its nosedives that analysts more politely call a downward cycle. The fallout of the first Gulf War, the spike in fuel prices, a generally lousy world economy, and very low yields in a deregulated U.S. industry all got their share of the blame.

The big breakthrough for the 777 had come in the fall of 1990 when United Airlines launched the plane with the PW4084. GE stunned the industry when its brand new GE90 won the next order from British Airways that had always been a Rolls-Royce customer. Pratt struck back winning orders from All Nippon, a traditional GE customer, and JAL that had rejected the PW4000 for its 747–400 only a few years previously. Then Rolls had its own coup when it won a huge 777 order for the Trent from Singapore Airlines that up until that time had never had anything but Pratt & Whitney engines. It was about as tough a competitive world as anyone could imagine, and profits on these sales seemed to be on a horizon that kept receding from all three companies.

With literally the whole aviation world watching, Pratt & Whitney came up with an extremely disciplined development plan for the PW4084. The team set up a system of "filters" for each step of the process from preliminary design to production engine. "The development program and all its associated problems enter at one end of the eight filters and pass through successive filters until production engines are ready for service entry," explained Tom Davenport, PW4000 program manager to *Aviation Week* in a 1994 interview.

The 112-inch fan PW4084 installed on United Airlines' first Boeing 777.

One of the key steps in this filtering process was going back and looking at every inflight shutdown that had occurred on a JT9D, PW2000, and 94-inch fan PW4000 in light of the requirement for 180-minute ETOPS. "We have a tremendous data base on our engines," Davenport said. "In doing this we determined what features we needed to implement to prevent a recurrence of the problem . . . and what tests were needed to verify the new design."

The most grueling test was perhaps a 3000-cycle program that ran from June through October 1994. The engine was run as close to ETOPS operating conditions as possible,

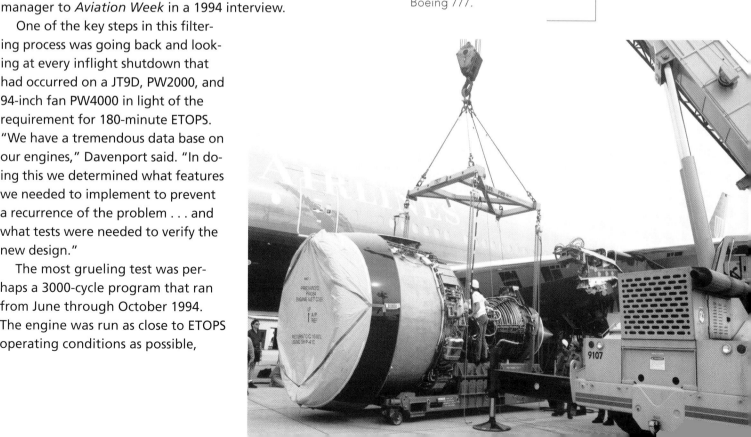

including simulating three diversion flights at full power. For part of the test program, the engine was deliberately unbalanced to see how it ran under high vibration stresses. And during the tests customer airline mechanics did routine maintenance as they would do in normal flight operations. In fact much of the engine design, especially externals, was guided by mechanics. Airline people had been invited to East Hartford early in the program to see how they would handle chores on a mock-up of the PW4084. Every time they found something they didn't like, they hung a red tag from it. After the first session, the Pratt & Whitney engineers said their engine looked like a Christmas tree.

Lessons learned included color coding as many items as possible to make it easy to trace lines and wire bundles, using captive fasteners, and even putting a step and handhold on the fan case so that mechanics would not grab a wire bundle to get up to the top of the case.

The details would get the PW4084 180-minute ETOPS out of the box. For instance, the analysis of all those inflight shutdowns showed that on some occasions the oil tank cap had not been replaced properly and oil leaked. So the cap and access door for the oil tank were redesigned so that you could not close and latch the access door if the cap were not secured properly.

The program was grueling in its rigor and at times nerve-racking in its schedule. United Airlines was putting the aircraft into service on a 180-minute ETOPS route between Washington and London in June 1995, and everything had better work. The first two flight-test engines had to be shipped on a giant Russian Antonov transport from Connecticut to Seattle. Mechanics put on the finishing touches during the flight.

The filter system had done its job, and the 777 flew for the first time on June 12, 1994. The 777 did 1000 cycles during the flight test, including eight 180-minute ETOPS diversions. The engines performed flawlessly. There was one somewhat embarrassing moment early in the program. The first flight-test engine had been mounted on RA001, the original 747, now a flying test bed. Taking off one day from Boeing Field, the engine surged with three loud bangs and a puff of smoke. The engine recovered within seconds, and a minor adjustment solved the problem. As luck would have it, however, a network TV crew happened to be shooting a story that day at Boeing Field on an upcoming Asia–Pacific economic summit. They got the whole thing on tape, and it became an issue for a short time.

The diligent work of Pratt & Whitney, Boeing, and customer airlines paid off on May 30, 1995. The FAA granted 180-minute ETOPS to the 777, the first plane in history to win that from its first day of service.[5]

---

[5] After demonstrating outstanding reliability in service, the 777/PW4084 combination was allowed a 207-minute ETOPS for certain Pacific Ocean routes.

# A New Company
## for the New Century

## chapter 11

**IN THE LATER YEARS OF THE 20TH CENTURY,** Pratt & Whitney would face some very difficult issues, as would its two big competitors and most of the aerospace industry. These companies' structures had been designed first to win World War II, and then to fight the Cold War, convert aviation to jet aircraft, support a worldwide air transport system, and put man into space and on the moon. Companies had sprawling facilities and lots of people. The cost of this entire infrastructure and the products it made was not a top priority. Governments were generous with research and development money, and there were lots of programs, military and civil, that helped build volume to keep all of those plants and people busy.

But even by the early 1980s, things were changing for Pratt & Whitney and other companies. Pratt & Whitney President Robert Carlson cautioned the UTC board of directors as early as 1982 that the company would have to change as aerospace matured. Airlines had lost $3 billion in the previous two years, military spending and programs were down, and the playing field for the big three in the engine

ACE and lean thinking brought new levels of quality and productivity to Pratt & Whitney as employee teams worked to figure out the best way to lay out the factory and improve performance throughout the company.

Not so strange bed-fellows after all. Pratt & Whitney and GE formed the Engine Alliance to build the GP7000 engine for the Airbus A380.

business was leveling off. "The (airline) customer base is already deteriorating and in several cases has balance sheets reflecting a negative net worth, " Carlson said. "It is unrealistic to think that any supplier will again have a technology edge that will be significant both in terms of marketplace and lasting in terms of time." He foresaw that Pratt & Whitney was in a fight with GE and Rolls-Royce in commercial and military as tough as it had ever faced. His concerns about the financial health of the airlines would echo time and again in the next 20 years as the industry would go from boom to bust at a dizzying clip.

Pratt & Whitney and most other aerospace companies would soldier along through the 1980s pretty much as they had for years, but the end of the Cold War in 1989–1990 and the turmoil caused by the first Gulf War would force the whole industry that had perhaps only paid lip service to costs and lean structure to get with it and fast.

By 1992 when the company was facing a $200 million-plus loss, something radical was needed. George David had just become president of UTC and would soon assume the chairman's title with the retirement of Bob Daniell. One of his first moves was to send Mark Coran, a veteran financial executive, back to Pratt & Whitney from UTC headquarters to do something about manufacturing. The story is told in great detail in James Womack's and Daniel T. Jones', book *Lean Thinking*, which argues that Pratt was the "acid test" of whether a huge, somewhat lumbering aerospace company could adopt the lean principles that had propelled so much of Japan's postwar resurgence.

Coran recalled: "Very suddenly, just when I arrived, everything that could go wrong went wrong. Rather than a simple cost-cutting exercise to deal with a 10 percent drop in volume, I realized that we needed to rethink the whole business." He would be helped in that effort by Karl Krapek, who was named Pratt & Whitney president in December 1992 with a mandate for major changes. Krapek and George David had both come from UTC's industrial companies, Otis and Carrier, and were very familiar with what had been going on in Japan. David convinced the Japanese manufacturing consultants Shingijutsu to come to work for UTC and Pratt & Whitney, beating out GE. "I was thrilled. We desperately needed their knowledge and snatched them away from GE at the last minute," David explained to Womack. They were experts in what came to be known as "one-piece flow" in manufacturing, the core of a lean operation. This does away with piles of expensive inventory and work in process under a "batch processing" system, very traditional in American factories. While working for Otis, David and Krapek had known Yuzuru Ito, who had been head of quality at Matsushita. They got him to come to UTC and established what became known as Ito University, where the soft-spoken guru would do relentless root-cause analysis of quality problems and teach others to do the same. It is safe to say that many people at Pratt & Whitney were not enthusiastic at first about all this help. There was a lot of pain for a lot of people, but it was clear the company would have to adapt to succeed in the new aerospace world of the 21st century.

"Our weekly output of three large engines and six small engines can literally be fitted in my office. So why do we need ten million square feet of manufacturing and warehousing space?" Krapek asked. Soon Pratt & Whitney plants in North Haven and Southington, Connecticut, would close with their work consolidated into East Hartford and Middletown. Some of that work would go to outside companies who were better at it and could do it at a lower cost. The real pain would come not just in the elimination of bricks and mortar but with the loss of 10,000 jobs in restructuring in the early 1990s.

One of the biggest physical changes in Pratt & Whitney would come in 1999 when the Government Engine Business would move lock, stock, and barrel from the sprawling West Palm Beach campus back into Connecticut facilities. Only outdoor test facilities and liquid rocket engine work would be left in Florida. It was traumatic for Florida and tough on the thousands of employees who would have to move. Familiarization trips were planned for employees and spouses to come up to Connecticut to look for houses and get a feel for the place. Some still laugh that on one of the first trips from Florida to Hartford the arriving employees were greeted by a snowstorm. It took a tremendous effort to make it all go smoothly over the next couple of years, but the move was accomplished without missing a beat on the F-22 and JSF programs.

The PW6000 for the Airbus A318.

Part of the reason for the move from Florida was a whole new structure for the company. Over the years Pratt & Whitney had become a collection of silos. Efforts had started in the PW2000, PW4000, and F119 programs to get people working cooperatively across functions with some success, but not nearly enough. Inside the company there was still a tendency toward "the Pratt Salute," pointing the blame to any other group or department but your own. Parts traveled miles back and forth across the company before finally getting put in an engine or shipped as spares. At one point Pratt had two kinds of expeditors, one for east–west aisles and one for north–south! And in engineering, administration, and other nonmanufacturing jobs the company was spending time and money on what then Commercial Engines President Sel Berson would bemoan as "white collar scrap, rework and repair."

The reorganization of the company created "module centers" in which everyone responsible for designing, developing, or making a major engine component worked together whether it was a military or commercial engine, whether they were engineers or manufacturing types. That transom between engineering and manufacturing finally was becoming an open door. There would be a center for the compressor section; one for combustors, augmentors, and nozzles; a turbine module center; one for engine controls and accessories; and one responsible for assembly and production testing.

And running the whole organization would be a new operating system dubbed Achieving Competitive Excellence or ACE. UTC and Pratt & Whitney, like almost every U.S. company going back to the 1980s, had tried several different approaches to quality with mixed results. With the advent of Shingijutsu and Mr. Ito in the 1990s, lean manufacturing and quality came together to formulate ACE. It is a detailed, disciplined system on how to run the company. What set ACE apart from other quality/productivity initiatives is that it is based on "market feedback analysis." Simply put, ACE starts with customers and how they rate the company. What are the problems and issues? How does Pratt & Whitney's "scorecard" look? Then there is an analysis of any gaps between what the customers want and what they are getting. The principle is that you then target improvements squarely at customer needs.

Soon everyone at Pratt knew their 6S's (sort, straighten, shine, standardize, sustain, and safety) or how to certify their process or define their standard work and how to benchmark and value stream map accurately. Not that there wasn't some moaning and groaning in long ACE sessions or that ACE coordinators weren't the victims of some bad jokes, but it did work. Another thing that made ACE different from other quality/cost initiatives was that it was here to stay. There would not be a new set of acronyms and letters to learn the next year. It might have been that factor that overcame the initial skepticism of yet "another program" and got people truly involved. And they were proving to themselves that it worked. Between 2005 and 2006 the parts of the company that had reached the highest levels in the program saw working capital turns improve 13 percent, revenue per employee grow 20 percent, and the scrap, rework, and repair rate improve 20 percent. Another telling statistic was that since ACE had really started earnings growth for all of Pratt & Whitney had been 20 percent, and the company sales had grown to well over $11 billion.

# NOT-SO-STRANGE BEDFELLOWS

Even when aviation is in turmoil, companies are looking at new programs. Boeing's 747 had been a big gamble, but a huge success. The plane came to symbolize global air travel. Airbus had grown rapidly and was ready to take on

The Geared Turbofan™ engine can bring a new level of environmental performance to commercial aviation.

Boeing's top dog, completing the Airbus journey to a full-line supplier. Boeing responded with plans for a stretched 747 beyond the 747-400. With the difficult financial situation incurred by the three engine makers on the A330 and the 777, the companies were somewhat leery of doing "three-on-a-wing" for these proposed aircraft. Out of this concern yet another joint venture between major engine companies emerged in 1996: The Engine Alliance was a 50/50 partnership between Pratt & Whitney and General Electric. The concept was to merge the best of the technologies from the 112-inch fan PW4000 and the GE90 to build a new engine, the GP7000, for the new generation of super jumbos. This limited the financial and technical risks for both companies and

The RD-180.

the customers in what appeared to be a somewhat narrow market. Rolls-Royce elected to develop yet another version of its Trent.[1]

The GP7000 would be in the 80,000-pound thrust class and use the core of the GE90 combined with the low-pressure section of the PW4000. The fragility of the market quickly became apparent when Boeing shelved its 747 growth plan. The Airbus entry, the A3XX, moved slowly for years, and the Engine Alliance seemed to be something of a stepchild for Pratt & Whitney and GE. But Airbus

---

[1] During this period, industry analysts kept saying there would be a consolidation in the engine business, speculating that Pratt & Whitney and Rolls-Royce might merge. Both companies pointed out that in effect the engine business had consolidated with the creation of joint ventures such as CFMI, IAE, and the EA plus the galaxy of other partners that surrounded each of the big three.

pressed ahead, and the A3XX became the A380. The GP7000 was off and running. As of this writing, it has leading market share on the big jet from Airbus. Delays unrelated to either engine snarled the A380 program, but the GP7000 is scheduled for entry into service in 2008.

While Boeing and Airbus were looking at the high end of the market, they were also gazing down below. The emergence of very successful regional jet programs by Bombardier and Embraer had interested the two big companies in the 100-passenger market. Boeing had new-generation 737s that got close and had inherited the MD95 program with the merger with McDonnell Douglas. The plane became the Boeing 717 although it never achieved significant market success. Airbus came to the conclusion that it had to protect the lower end of its A320 family and shrunk the A319 into what became the A318.

> **"Even when aviation is in turmoil, companies are looking at new programs."**

The engine for this new plane became a bit confusing. CFMI offered a lower thrust version of its CFM56, which met the criteria, but IAE was reluctant to come up with a derated V2500. Pratt & Whitney and Rolls-Royce believed that the weight and drag penalties would have made such an engine unattractive. Pratt & Whitney also had an issue with its stable of engine cores. Nothing had replaced the JT8D in the 18,000–24,000-pound thrust class. The V2500 was a 25,000-pound thrust core, and Pratt & Whitney Canada was only just getting up into the 10,000-pound and above class. So out of the need for a new core and the A318 program came the PW6000.

For many years fuel burn had been the driver in engine design, with companies straining every technical muscle they had to get a percent improvement here and a percent improvement there. But the 100-passenger market demanded something different. These airplanes would do a lot of daily flights on relatively short routes. Those percentage-point fuel burn improvements only really paid off for airlines on long-haul routes. With several takeoffs and landings every day and a relatively short time at economical cruise power, fuel burn improvement was not the major driver. Durability and maintenance costs were more important for the 100-passenger market. The design of the PW6000 eschewed exotic materials and went for ruggedness and fewer parts. This led to a very ambitious five-stage high-pressure compressor. Fewer stages did indeed mean fewer parts and less maintenance, but it also meant that each stage had to do more work. In flight test the engine worked perfectly except that the compressor could not give acceptable fuel burn, rugged as it was. As sometimes happens in aviation, the effort to stretch for one goal hurt someplace else. Pratt & Whitney partner MTU came up with a design for a six-stage high compressor that was still very robust with low maintenance costs and got the right fuel burn. Also unique to the PW6000 is the fact that MTU does final assembly in Munich. It is the first Pratt & Whitney engine, other than wartime models, that is assembled outside the company.

In engineering, every problem should be an opportunity to learn something, and so it was with the PW6000. The ambitious compressor design had relied to a large degree on computer simulation, but there had not been technology readiness and demonstration programs with hardware. Pratt & Whitney President Louis Chênevert pointed out at the time that a technology demonstrator program could have avoided problems. In the future "technology readiness" would be standard operating procedure. This would be combined with a detailed engineering standard work discipline to make sure programs hit their targets. Standard work is a key part of the whole ACE structure. Basically standard work means that every repeatable process is consistent and waste-free using the best possible practices that teams develop. These can be shared across the whole company. You could say standard work means you do your best every time exactly the same way with the minimum resources required but with maximum results.

# AN ENGINE FOR GEAR HEADS

And so it is with the Pratt & Whitney's new commercial engine for the next generation of aircraft, the Geared Turbofan™ (GTF) engine. The concept goes a long way back, in fact to propellers. They are very efficient ways of powering an airplane but run into stability problems at higher and higher power when the blade tips approach supersonic speed. You could say that high bypass turbofans with that big fan upfront combine the very good features of a propeller with the power of a jet. Still the fan can only turn so fast or it will lose performance. The compressor and turbine can stand much higher speeds and thus generate more power, but the fan cannot. Designers always had to strike a compromise when the core engine was coupled directly to the fan as it is in the standard twin-spool jet design for Pratt & Whitney and GE. Rolls-Royce started to address the problem with a three-spool design to get better fan efficiency. There is an unending debate between engineers as to which approach works best. In the early 1980s another concept entirely emerged. Pratt & Whitney and Hamilton Standard called theirs a prop fan. GE called its the unducted fan or UDF. The idea is to have a huge, many-bladed propeller-like device on the outside of the engine like a traditional propeller instead of a fan inside the case. You run the core engine at optimum speeds and use a gearbox, nominally with a three-to-one reduction, to get that down to a speed that runs the huge prop fan efficiently. You can then design a very large fan with a high bypass ratio and thus great efficiency, fuel burn, and low noise as compared with standard turbofans. Both companies flew demonstrators on MD-80 aircraft, but the airlines never went for them. The idea of having a huge scimitar-shaped prop with a dozen blades spinning in full view of the passengers was daunting. Prop Fan and UDF

faded, but from time to time the idea re-emerges. Giant supersonic props also impose a noise penalty even as they achieve better emissions performance. Installation of such a large structure can also pose a set of daunting problems. Reliability is another issue. Industry data show that pitch change mechanisms required for a big prop fan have higher maintenance requirements than gearboxes, despite some conventional wisdom that gearboxes are difficult. But Pratt engineers knew they were on to something even if the prop fan concept was not a winner.

One of the fascinating parts of the story of the GTF engine was how to cool a gearbox that would have to absorb thousands of more horsepower than what existed in traditional turboprops. Engineers literally got down and dirty with gearboxes and made an interesting discovery. In gearboxes most of the cooling oil simply sloshed around and really did not do all that much cooling. So they worked on ways to precisely deliver the cooling oil to the teeth of the gears. They became experts at watching high-speed photography of tiny oil droplets hitting gear teeth. Also they figured out ways of scavenging the oil out of the gearbox quickly. This meant that you needed less oil, and it would not be exposed to high heat for

It's a long way from Capitol Avenue and a tobacco warehouse. The 2007 groundbreaking for an engine overhaul center with China Eastern Airlines.

so long. So you needed less cooling and a relatively small oil cooler, which had been concerns of airline operations people when looking at the prop fan or UDF.

Also, realizing the reluctance of airlines to go for a big fan outside the engine, they first came up with the advanced ducted propulsor, a big fan inside the engine case. In 1992 a demonstrator ran successfully using a standard PW2000 engine core with a 40,000-horsepower gearbox and a pitch-change mechanism for the 118-inch fan. Although technically impressive, it was not enough to convince airlines to take that big a leap while the standard turbofan was quite satisfactory. Pratt & Whitney, however, stuck with the work. Given 20 years of research and development work to the tune of several hundred million dollars, Pratt & Whitney engineers came up with the Geared Turbofan™ engine, an ultra-high bypass fan, but contained in the fan case and without variable pitch. In aircraft up to 200 passengers, a geared turbofan engine reduces fuel burn, and thus emissions, by 12 percent over traditional designs, while cutting noise by 20 decibels below the Stage 4 noise regulations. The noise level is about half of traditional high bypass turbofans. These were powerful arguments for adopting geared fan engine technology in an era when airlines were coming under increasing environmental pressure and fuel costs were a huge concern.

The program is expected to be a vital part of Pratt & Whitney's future and targets the next generation of single-aisle aircraft. Business considerations had convinced the company not to participate in the Boeing 787 or Airbus A350 wide-body programs and to focus instead on the new single-aisle market, planes to replace the 737 and A320 families. Market studies indicated that the GTF engine should initially target a 20,000- to 30,000-pound thrust range, but the technology could move up or down and will be part of Pratt & Whitney Canada's future as well.

"The GTF engine has the advantages our customers are looking for," said Bob Saia, head of commercial engine development programs. "By using the speed reduction gearbox between the engine's low-pressure turbine and the fan, each component is able to rotate at its optimum speed. You get the benefit of more power in fewer stages, with fewer parts, lower costs, and lighter weight."

All of the hard work came to fruition in the fall of 2007 when Mitsubishi Heavy Industries announced it would build a new technology 70-to-90 passenger regional jet and it would be powered by a Pratt & Whitney Geared Turbofan engine. A key advantage was that the GTF could meet Mitsubishi's schedule. The Mitsubishi Regional Jet program was officially launched on March 28, 2008, with an order from All Nippon Airways. Bombardier also selected the Pratt & Whitney Geared Turbofan engine for its CSeries aircraft. Both aircrafts are scheduled to enter into service in 2013.

Other approaches to reductions in the environmental signature of commercial engines were years and years out in the future or involved relatively small improvements that would not make significant environmental or economic gains. The GTF is set to be an integral part of Pratt & Whitney's commercial engine business for many years to come.

# YOU DO HAVE TO BE A ROCKET SCIENTIST

As early as the mid-1950s, Pratt & Whitney recognized that it would have to get into rockets to survive in an industry that was changing from aviation to aerospace. Earlier we detailed these ventures and their success, although the company did not achieve the scale it had hoped.

The very successful RL10 liquid-hydrogen engine had kept the company's hand well into space propulsion, and in the 1990s this got a boost by hooking up with the Russians—yes, the Russians. Russian booster engines were always a technology star in the Soviet Union, and with the fall of the Soviet regime, that technology became available to the West. Pratt & Whitney made overtures and eventually created RD-AMROSS, a joint venture with the Russians who were looking for a way to keep their design bureaus active in the postcommunist era. The product that emerged from the joint venture was the RD-180, a million-pound thrust rocket engine that was 10 percent more powerful than anything the United States had developed. It became the booster engine for the Lockheed Martin Atlas III and Atlas V launch vehicles.

> **In August 2005, Pratt & Whitney Rocketdyne formally came into existence.**

Photo courtesy of NASA.

Pratt & Whitney Rocketdyne engines launch the space shuttle.

**133**

As the new century got underway, it became apparent that the U.S. rocket industry would have to consolidate. Both military and commercial launch programs were fewer, and the market was tight. The open-ended budgets and programs of the Cold War and the race to the moon were long gone. One of the pioneers in space propulsion was Rocketdyne, which had started life as part of North American Aviation and became part of Rockwell and finally Boeing. In 2005 Boeing decided that building rocket engines was not a core business and began looking for a buyer. Propulsion was the core business of Pratt & Whitney, and the company wanted to enhance its business in space. So a deal was struck, and in August 2005 Pratt & Whitney Rocketdyne formally came into existence.

The history of Rocketdyne is a fascinating story.[2] At the end of World War II, Dutch Kindelberger and Lee Atwood, the heads of North American Aviation, had

---

[2] Readers interested in a detailed account should seek out *Rocketdyne: Powering Humans into Space* by Robert S. Kraemer, published by the AIAA in 2006.

Image courtesy of United Launch Alliance.

Delta IV Heavy DSP-23 launch.

taken note of the success German scientists had had developing rockets such as the V2. They believed that North American had better start figuring out this new rocket business as part of its future. A small team of young engineers was put together and began fiddling with liquid-fueled engines. Their first "test site" was part of the North American parking lot at its Inglewood, California, facility. Fortunately, there were some blockhouses on the property put there during the war in case of a Japanese air attack. One got plunked down in the middle of the new test area/parking lot, and the team started working with some surplus thrust chambers loaned by Jet Propulsion Laboratory and later some surplus jet-assisted takeoff motors (JATO). The first motors from JPL were only 50 pounds thrust, and team members recalled that they only whistled rather than roared.

A breakthrough came when the fledging operation won a contract to build engines for the Navaho missile. Company historian Robert Kraemer also credits the early decision to associate with Wernher von Braun and his Army/NASA team at the Redstone Arsenal in Alabama with making Rocketdyne what it became. Rocketdyne engines would power the first U.S. satellite launch and the Redstone missile that made Alan Shepherd the first American in space. A Rocketdyne engine took John Glenn on the first orbital flight. Rocketdyne built the motors for the Jupiter, Saturn, Saturn I, Saturn IB, Saturn V, and the space shuttle (remember they beat Pratt & Whitney for that job). Pratt & Whitney Rocketdyne has begun the research for a new generation of space vehicles, like the Ares I, that could put Americans back on the moon. It is also looking ahead to systems for in-space propulsion and hypersonics. The acquisition of Rocketdyne makes Pratt & Whitney a complete propulsion company from the little Pratt & Whitney Canada PT6 to the million-pound thrust RD180. It now has the broadest technology base of any company in aerospace propulsion.

> 66 Pratt & Whitney Rocketdyne has begun research . . . that could put Americans back on the moon. 99

# AT YOUR SERVICE— AROUND THE WORLD

Beginning in the 1990s, Pratt & Whitney went through a profound change, another reinvention of its business. As in the past, the focus was on giving customers the best engines, but that vision now looked more than ever at what happened after the engine left the factory. It is no secret that for decades Pratt & Whitney and the other engine companies made their money off spare parts. They did little service themselves. Pratt & Whitney's maintenance, repair, and overhaul business (MRO) was essentially a warranty shop. The practice was to set up airline customers with their own overhaul shop and then help them become third-party providers themselves. It was a way to sell engines. Buy our engines, and we will put you in the overhaul business. It was not unheard of for sales people to suggest to the overhaul people to back off a deal that might hurt an airline shop's business. For many years the system of selling new engines and spare parts and doing only very limited overhaul and repair work did just fine. Change, however, was coming.

Perhaps most importantly, airlines began looking at their own businesses. It turned out many did not want to cook meals, run hotels, or even fix engines and airplanes. They wanted to sell tickets and fly. Some of the most financially successful airlines outsourced their maintenance. This was particularly true if an airline bought a new aircraft/engine combination. Setting up a shop or shops to take care of it could get very expensive. In many cases part of an engine sale now was not setting up a shop for an airline, but agreeing to a long-term maintenance deal as part of the sale.

Even air forces were looking for help. Some countries did not have the technical knowledge base to maintain increasingly advanced equipment. In other cases budget cuts had cut the number of personnel available so that it was better to

Mitsubishi chooses the Pratt & Whitney Geared Turbofan™ engine for its new regional jet.

use outside contractors rather than tie up uniformed or civilian personnel. A simple example is the Army hiring caterers to feed the troops in the field.

All of these trends gave Pratt & Whitney an opportunity to leverage its technical expertise and transform itself once again. The company built a network of more than 20 overhaul and parts repair centers all over the world for large commercial and military engines, Global Service Partners. And it is not just wrench turning. Pratt & Whitney offered complete engine fleet maintenance programs, material requirements planning programs, real-time performance data collection and analysis, and even line maintenance. An advantage the company brought to this market was its engineering expertise. The same engineers who designed and tested engines invested millions in developing new, innovative parts repairs that save customers money and improve performance. And in the 21st century this emphasis also means greener engines. For instance, Pratt developed a service called EcoPower® wash. It is a completely closed-loop engine wash system that uses plain water, no chemicals, no runoff. The work can be done with the engine on-wing, and a cleaner engine burns less fuel and produces less emissions. The Pratt network can handle CFMI and Rolls-Royce commercial engines as well as Pratt & Whitney engines. The company even manufactures new parts for the CFMI engine as well as offering parts repair. In 2007 Pratt & Whitney came up with its own new acronym to describe itself. It's not an OEM (original equipment manufacturer). It's not an MRO (maintenance, repair, and overhaul). It's an OEMRO!

A key to this effort has been partnering with airlines around the world in establishing a high-technology service business. They bring a whole new set of insights to providing the best possible service. These include Singapore Airlines, Air New Zealand, China Eastern Airlines, and THY Turkish Airlines. The Norway Engine Center, which works on the CFM56, for instance, was originally established by Braathens and then purchased by Pratt.

This new business profile and Pratt & Whitney's growing presence in manufacturing facilities and service centers all over the world, not just in North America, are proving just as transforming as the Wasp or the move from piston engines to jets. The next chapter in Pratt & Whitney's history of dependable engines is being written.

# BIBLIOGRAPHY

## Books

Drewes, Robert W., *The Air Force and the Great Engine War*, 1987, National Defense University Press, Washington, DC.

Fernandez, Ronald, *Excess Profits: The Rise of United Technologies*, 1983, Addison-Wesley Publishing, Reading, MA.

Garvin, Robert V., *Starting Something Big: The Commercial Emergence of GE Aircraft Engines, 1998*, American Institute of Aeronautics and Astronautics, Reston, VA.

Gunston, William, *World Encyclopedia of Aero Engines*, 4th Edition, 1998, Patrick Stephens Limited, Sparkford, Somerset.

Kraemer, Robert S., *Rocketdyne: Powering Humans Into Space*, 2006, American Institute of Aeronautics and Astronautics, Reston, VA.

Mead, Carey, *Wings Over the World: The Life of George Jackson Mead*, 1971, The Swannet Press, Wauwatosa, WI.

Monday, David, editor, *The International Encyclopedia of Aviation*, 1977, Crown, New York.

Mulready, Richard, *Advanced Engine Development at Pratt & Whitney*, 2000, Society of Automotive Engineers, Warrendale, PA.

Rentschler, Frederick B., *An Account of the Pratt & Whitney Aircraft Company, 1925-1950*, 1950, privately published.

Rowe, Brian H., *The Power to Fly: An Engineer's Life*; 2005, American Institute of Aeronautics and Astronautics, Reston, VA.

Sullivan, Kenneth H. and Milberry, Larry, *Power: The Pratt & Whitney Canada Story*, 1989, CANAV Books, Toronto.

White, Graham, *R-2800: Pratt & Whitney's Dependable Masterpiece*, 2001, Society of Automotive Engineers, Warrendale, PA.

White, Graham, *R-4360: Pratt & Whitney's Major Miracle*, 2006, Specialty Press, North Branch, MN.

Wilson, Eugene E., *Kitty Hawk to Sputnik to Polaris*; 2nd Edition, 1967, Literary Investors Guild, Palm Beach, FL.

Wilson, Eugene E., *Slipstream*, 1950, McGraw-Hill, New York.

Womack, James P. and Jones, Daniel T., *Lean Thinking: Banish Waste and Create Wealth in Your Corporation*, 1996, Simon and Schuster, New York.

## Major Company Publications

Bobbi, Mark A., Manager Employee Education, *Pratt & Whitney: A 70th Anniversary Tribute, 1995*, United Technologies Corporation.

*Partners in Defense: Pratt & Whitney's First 25 Years in Florida*, 1983, United Technologies Corporation.

*The Pratt & Whitney Aircraft Story*, 1950, United Aircraft Corporation.

*Pratt & Whitney 65th Anniversary: In the Company of Eagles*, 1990, United Technologies Corporation.


## Archival Material

*Aircraft Nuclear Propulsion Program of Pratt & Whitney Aircraft Division*, 1980, Archive and Historical Resource Center, United Technologies Corporation.

*Bye Bye Blackbird: Retirement of the SR-71*, 1990, Archive and Historical Resource Center, United Technologies Corporation.

Crow, David E., *Address to Royal Aeronautical Society*, 2000, Archive and Historical Resource Center, United Technologies Corporation.

*The F100 Engine*, 1980, Society of Automotive Engineers Warrendale, PA.

*F-14A Status Report*, 1974, Society of Automotive Engineers Warrendale, PA.

*The Historical and Technological Development of the JT9D*, 1982, Archive and Historical Resource Center, United Technologies Corporation.

*A History of Pratt & Whitney Aircraft's Florida Research and Development Center and Government Products Division*, 1982, Archive and Historical Resource Center, United Technologies Corporation.

*The History of United Technologies Chemical Systems Division*, 1983, Archive and Historical Resource Center, United Technologies Corporation.

*Interviews with Charles Roekle, Richard Mulready, and Frank McAbee,* Archive and Historical Resource Center, United Technologies Corporation, interviews conducted 1982.

*Interviews with H. M. Horner, 1945–1968*, Archive and Historical Resource Center, United Technologies Corporation, interviews conducted 1960, 1973.

Slay, Gen. Alton D., *Presentation to U. S. Senate Armed Service Committee,* 1979, Archive and Historical Resource Center, United Technologies Corporation.

*Statement of United Technologies Corporation–Pratt & Whitney*, U. S. Senate Defense Appropriations Subcommittee, 1982, Archive and Historical Research Center, United Technologies Corporation.


## Periodicals and Newspapers

"ADP Demonstrator Finishes First Sea-Level Tests," *Aerospace Propulsion*, Vol. 3, No. 24, Nov. 26, 1992, p. 1.

"Aircraft Engine Program Sparks Fierce Lobbying," *The Washington Post*, March 12, 1979, p. A1.

"Airlines Could Reap Big Savings with New Pratt Common Core," *Aviation Week and Space Technology*, Vol. 136, No. 23, June 8, 1992, p. 43.

"C-17 Go-Ahead Secures Future of P&W's PW2000 Series," *Aerospace Propulsion*, Vol. 6, No. 23, Nov. 9, 1995, p. 1.

"Cheap is Good," *Air Transport World*, Vol. 35, No. 8, Aug. 1998, p. 50.

"Commonality Key to PW4084 Early ETOPS," *Aviation Week and Space Technology*, Vol. 140, No. 15, April 11, 1994, p. 53.

"Engine Slowdown," *The Hartford Courant*, Feb. 22, 2002, p. E1.

"F119 Design to Slash Maintenance Requirements," *Aviation Week and Space Technology*, Vol. 143, No. 4, July 24, 1995, p. 56.

"F119 Versatility Challenged by JSF Requirements," *Aviation Week and Space Technology*, Vol. 145, No. 22, Nov. 26, 1996, p. 25.

"F-22 Raptor Meets First-Flight Goals," *Aviation Week and Space Technology*, Vol. 147, No. 11, Sept. 15, 1997, p. 22.

Fink, David, "The JT8D Engine Uses Full-Length Fan Duct," *Aviation Week and Space Technology*, Oct. 14, 1963, p. 52.

"First RD180 Delivered," *Aviation Week and Spaced Technology*," March 25, 1998.

"German HP Compressor Set For Two New Pratt Engines," *Aviation Week and Space Technology*, Vol. 153, No. 9, May 27, 2002, p. 56.

"The Great Engine War," *Northeast Magazine*, March 27, 1983, p. 10.

"IAE Cites Risk, Scuttles Plans To Launch Superfan," *The Hartford Courant*, April 9, 1987, p. C32.

"McDonnell Douglas's F-15 Fighter Appears Headed for Big Air Force Production Run," *The Wall Street Journal*, Feb. 12, 1973.

"New Engines Launched," *Aviation Week and Space Technology*, Vol. 142, No. 8, Feb. 20, 1995, p. 31.

"New Jumbo Jet Engines Are Planned By Pratt," *The New York Times*, Dec. 8, 1982.

"P&W Gears Up Family Plan as PW9000 Goes On Line," *Flight International*, April 3–9, 2007, p. 7.

"P&W to Lead Consortium in Development of Airbus Engine," *The Hartford Courant*, June 11, 1987, p. B1.

"P&WA Offering Engine Warranty," *The Hartford Courant*, Aug. 4, 1981, p. C1.

"Pratt & Whitney Aims for Even More Power from the F135 Engine for the Joint Strike Fighter," *Aviation Week and Space Technology*, Vol. 168, No. 5, Jan. 17, 2005.

"Pratt & Whitney Expects to Regain Market Share with PW2000, PW4000 Engines," *Aviation Week and Space Technology*, Aug. 31, 1987, p. 70.

"Pratt & Whitney Preparing to Move Forward with F135 Engine Development," *Defense Daily International*, Vol. 3, No. 1. Nov. 2, 2001.

"Pratt & Whitney to Certify New Powerplants," *Aviation Week and Space Technology*, Dec. 13, 1982, p. 24.

"Pratt & Whitney's Surprise Leap," *Interavia Business and Technology*, Vol. 53, June 1998, p. 25.

"Pratt Achieves Strategic Victory With New Engine," *The Hartford Courant*, Sept. 8, 1998, p. A1.

"Pratt Begins Tests of JSF Powerplants," *Aviation Week and Space Technology*, Vol. 148, No. 25, June 22, 1998. P. 34.

"Pratt Claims Record Turbine Temps in JSF Engine Tests," *Aviation Week and Space Technology*, Vol. 154, No. 25, June 18, 2001, p. 76.

"Pratt Cries Foul in Race for $10 Billion Engine Business," *Defense Week*," June 1, 1982, p. 3.

"Pratt Opts for Increased Effort in Large Engine Market," *Air Transport World*, Vol. 20, Jan. 1983, p. 47.

"Pratt's ATF Engine Victory Could Yield 1500 Powerplants," *Aviation Week and Space Technology*, Vol. 134, No. 17, April 29, 1991, p. 24.

"Propfans To Offer New Generation of Power," *The Hartford Courant Business Weekly*, Sept. 8, 1986, p. 7.

"PW2037 Starts Work," *Flight International*, Nov. 10, 1984, p. 1246.

"PW4000 Off to a Good Start in Big Engine Race," *Air Transport World*, Vol. 24, March 1987, p. 32.

"PW6000 Engine Design Targets Costs, Reliability," *Aviation Week and Space Technology*, Vol. 150, No. 23, June 7, 1999. P. 44.

"Quick Start: Propulsion Development for the JSF is on Track and Progressing Rapidly," *Aviation Week and Spaced Technology*, Vol. 160, No. 16, April 20, 2004, p. 32.

"Risk, Cost Sway Airframe, Engine Choices for ATF," *Aviation Week and Space Technology*, Vol. 134, No. 17, April 29, 1991, p. 20.

"Rolls-Royce Leaves JT10D Turbofan Development Program," *Aviation Week and Space Technology*, May 16, 1977, p. 17.

"Russian Government Approves Energomash–UTC Deal," Reuters, April 1, 1996.

"Superfan Opponents Dispute Timing; UDF Tests Advance," *Air Transport World*, Vol. 24, No. 4, April, 1987, p. 42.

"Technical Snag Plagues Pratt & Whitney Jet Engine Project," *Palm Beach Post*, Nov. 27, 1970, p 1.

"2037: Pratt & Whitney Tests New Turbofan," *Flight International*, Dec. 19, 1981, p. 1832.

"United Technologies, Rolls-Royce Scrap Jet-Engine Plan as Needs Keep Varying," *The Wall Street Journal*, May 10, 1977, p 13.

"The V2500 Engine," *Exxon Air World*, Vol. 37, No. 3, 1985, p. 16.

"Vertical Reality," *Flight International*, April 20, 2004, p. 50.

Yafee, Michael L., "Pratt & Whitney Builds on JT9D Concept," *Aviation Week and Space Technology*, June 24, 1974, p. 49.

# Where Did the
## Eagle Come From?

**IT MAY BE ONE OF THE BEST-KNOWN LOGOS** in the world, and it has been on every Pratt & Whitney engine ever built. But no one knows how it got started. The late Harvey Lippincott, UTC's corporate historian and archivist, spent years combing every company record, book, library, and museum he could find to get at the origins of the Pratt eagle. It was to no avail. There is no record of who picked the American bald eagle as the company symbol, why, or who designed it. Frederick Rentschler makes no mention of it in his 1950 personal memoir.

The eagle, however, has been a hit since 1925. Many, many eagle medallions have quickly disappeared from engines and wound up on belt buckles, desk sets, tool boxes, car steering wheels, you name it. There have been a couple of variations on the standard enamel four-color eagle. The RL10 engine gets so cold that the enamel flakes off, and so a stainless-steel logo with no paint was created. It's the cryogenic eagle. In contrast, the J58 ran so hot that the enamel burned off. So the cryogenic eagle also became the high-temperature eagle also known as the Florida eagle because it was made in West Palm Beach. Those eagles pretty much stayed with the engines.

In 1981 the original eagle was replaced with a modern design. Frankly, most people disliked it. Within a few years the traditional eagle returned. Then company President Art Wegner was applauded as he walked through the factory on the day he unveiled the new, old eagle. Posters were distributed to virtually every employee depicting a photo of a very aggressive looking bald eagle with the caption, "I'm back. Wiser, Stronger, Tougher."

Wording on the medallion has changed over the years. The original read "Pratt & Whitney, U.S.A., Dependable Engines." In 1941 "Reg. U.S. Pat Off." was added and later dropped. During the war years, licensees used the regular logo, but Ford, Buick, Continental, or Nash-Kelvinator were on the eagle's blue background. The word "aircraft" was added in 1945 but dropped many years later.

# AIRCRAFT AND ROCKET ENGINES

| ENGINE MODEL | AIRCRAFT | NOTES |
|---|---|---|
| **CLASSIC ENGINES** | | |
| **R1340 WASP** | Amphibions N-2-C, Atlantic C-5, Bell H-12, Bellanca L-11 Skyrocker, AT-15BL, and JE-1; Boeing AT-15BO Crewmaker, P-26, P-29, C-73, (Model 247), F2B, F3B, F4B, Model 40A, Model 100 SP, Model 204; Buhl CA-6W; Canadian Car & Foundry C-64 Norseman; Cessna C-106A, XPG-4; Commonwealth of Australia Chase Wirraway; Vultee YBC-3; Curtiss Wright XA-4, XO-12 Falcon, P-3AHawk, O-52 Owl, SOC Seagull, A6A, Model 6000A; Detroit C-23 Al-tair, C-25 Altair; Detroit DHC-3, U-1; DeHavilland Otter; Douglas C-29 Dolphin 8, Model EJ-2, Model 25-2, BT-2, C-29, O-32; Eastern RD-1, 2, 3, 4; Fairchild F-1 Model 71, FB-3, C-96 Model 71, AT-13 Yankee Doodle, FC-2W; Fiat G-49; Fokker F-10A, F-22, S-13; Ford C-4, 5AT; Frye F-1 Safari; Goodyear K & M Ships; Grumman G-73 Mallard; Hamilton H-45; Junkers JU-52; Kaman HOK-1, 2; Laird LC-RW-450; Lockheed C-12 Vega, C-17 Vega, 10, UC-36B Electra Lockheed, UC-85 Orion, C-101 Vega, Altair, AirExpress 3, Executive; Macchi MB-323; McDonnell AT-15MC Crewmaker; Metal Flamingo; N.A.F. SON-1; Noorduyn AT-6 Harvard, C-64 Norseman; NorthAmerican AT-6 (SNJ) Texan, Harvard, BC-1, 2; Y1BT-10; Fokker Super Iniversal, North American NJ-1 Yale, NA-16; Northrop A-17AS Nomad, C-19, Pioneer, Delta; Piaggoio P-150; Pi-asecki HRP-1, 2 Rescuer; Ryan B-7; Sikorsky C-6, S-38, S-55 (HO4S); Stearman Model 4E, Alpha 4A, YOSS-1; Stinson SM-6B; Thomas Morse XP-13A, O-19, O-22; Vought OSU-1, XF2U-1; Corsair Vought O2U, O3U; Corsair Zenith Z-6-A. | A nine-cylinder, air-cooled recipro-cating engine originally producing 410 horsepower, later up to 600 horsepower. |
| **R-1690 HORNET** | American Airplane & Engine 100A Pilgrim; Atlantic XLB-2; Bellanca C-27; Boeing YB-9, Model 40-B, Model 80-A1, Model 95, Model 220, 221, Monomail, Y1C-18; Consolidated Model 17, 20, 20A Fleetster, Model 16 Commodore; Curtiss YA-10; Douglas O-38; Faucett F-19; Focke-Wulf FW-200 Condor; General GA-43, C-14B; Hamilton TG-1; Junkers JU-52, JU-86; Keystone (Loening) C2H, B-4A, LB-3A, LB-7, 8, 10A, 12, 13, 14; Lockheed C-56A, C, D Lodester, C-59 Lodestar; Lockheed XR40-1 Hudson; Lockheed Electra Model 14H2; Martin B-12, YB-13, P3M-2, T4M-1, Model 74; N.A.F. T4N-1; Sikorsky OA-8, Y10A-8, S-40A, S-41B, S-42, S-43 (JRS-1); Thomas Morse YO-20; Vought V-80P, V-92S; Corsair Vought SU-1,2,3,4; O3U-2. | A nine-cylinder engine, rated at 525 horsepower, with later models reaching 875 horsepower. A larger version, the R-1860, produced 575 horsepower. |

| ENGINE MODEL | AIRCRAFT | NOTES |
|---|---|---|
| | **CLASSIC ENGINES** | |
| R-985 WASP JR. | Airspeed Oxford V; Avion Max Hoiste MH-1521; Avro Anson V; Barkley Grow T8P-1; Kansas Beech F-2; Beech AT-7 Navigator; Kansas Beech AT-11 (SNB); Beech C-43 (GB) Traveller, C-45 (JRB) Voy-ager, C-45A-H Expeditor; D-17S, 18S Series; Bell Model 42, Model 200 (XV-3); Bellanca 300W Pace-maker, 400W Skyrocket; Bendix Model J; B/J OJ-2; Boeing XBT-17; Bristol Mk171; Canadian Fairchild F11X Husky; Beech CQ-3; Consolidated BT-7, Y1PT12; Vultee BT-13 (SNV) Valliant; Consolidated Model 21-C; Beaver DeHavilland L-20, DHC-2; Douglas Dolphin 3, OA-4, YOA-5, C-26; Fairchild Sekani, Model 51A; Fleetwings BT-12 Sophmore; Fletcher PO-11, YCQ-1A: Grumman OA-9 Goose, OA-13 (JRF), XJ3F-1; Howard UC-70 Nightingale, DGA-15; Koolhoven FK-50, FK-51; Lockheed UC-36, 37, 40 Electra; JO-1, 2, 3; XR20-1, R30-2; Model 12; Orion 2D; McDonnell XHJH-1; N.A.F. OS2N Kingfisher; NorthAmerican BT-14 Yale; Piasecki HUP-2, H-25; Platt Le Page XR-1; Ryan YO-51 Dragonfly; Seversky BT-8; Sikorsky H-5F, R-5 (HO2S,3S), C-28, S-39, S-51; Spartan UC-71 Executive, Model C5-301, Model 7W; Stearman BT-5, Model 4D, A75 N1; Reliant Stinson L-12A, UC-81; Stinson W; Vidan Research XBT-16; Vought OS2U Kingfisher; Waco UC-72 (SRE), S3HD. | A smaller, nine-cylinder recipro-cating engine designed for light transports, trainers, sports aircraft and helicopters. Initial rating was 300 horsepower, advanced models up to 600. |
| R1535 TWIN WASP JR. | Potez Model 63-12; Boeing Model 247A; Breguet Model 695-AB2, Model 699-B2, Model 690-C; Bristol Bolingbroke; Canadian Car & Foundry Model 10; Curtiss Wright SBC-3 Cleveland; Douglas O-46A; Fokker G-1; Great Lakes BG-1; Grumman F2F, F3F-1, XSF-2; Miles Master III; NorthropA-17, 17A Nomad, BT-1; Vought V-142, V-143, SBU-1, 2, 3; SB2U Vindicator, XSB3U. | The R1535 was Pratt & Whitney's first two-row production engine. A smaller version of the R-1830 Twin Wasp, it produced up to 825 horsepower. |
| R-1830 TWIN WASP | Aero-Sud Est SE-161; Bloch Model 153-C1, Model 174, Model 176; Boeing XB-15 (XC-105), PB2B Catalina; Breda BR-65; Bristol Beaufort II; Budd RB-1, 2 Conestoga; Canadian Car & Foundry CBY-3; Common-wealth ofAustralia Boomerang; Consolidated Vultee F-7 Liberator, OA-10 Catalina; Consolidated Vultee YA-19; AT-22 Liberator, B-24 (PB4Y-1) Liberador; Vanguard Vultee P-66; Liberator Express C-87 (RY-3), Consol-idated Catalina PBY, PB2Y Coronado; Consolidated Vultee PB4Y-2 Privateer; Model 39; Curtiss Wright P-36 Mohawk, XP-42, C-76 Caravan, H-81A, H-75; Douglas DSTA, DC3A (C-41), C-47 (R4D) Dakota Skytrain, C-48 Dakota Skytrooper, C-52 Dakota Skytrooper, TBD Devastator, C-53 (R4D-3, 4) Dakota Skytrooper, C-68; East-ern FM-1 Wildcat V; Fiat G-212CP Monterosa; Ford C-109 Flying Tanker; Grumman JF-1, | A two-row, 14-cylinder radial design delivering up to 1,350 horsepower. Production totaled 173,618 engines, more than any other aircraft engine. |

| ENGINE MODEL | AIRCRAFT | NOTES |
|---|---|---|
| **CLASSIC ENGINES** | | |
| **R-2800 DOUBLE WASP** | Aero Nord Model 2501, Model 2503; Aero Sud-Oest SO-30P; Bell HSL-1 (Model 61); Breguet BR-763, BR-765; Brewster XA-32, F3A-1 Corsair; Chase C-123B; Vultee XA-19B; Consolidated Vultee TBY-2 Seawolf, Model 110, Model 240; Model 340(R4Y) Convair-Liner; Model 440 Metropolitan, T-29, C-131 Samaritan, C-46 (R5C) Commando; Curtiss Wright P-60A, E, XC-113, XF15C, SB2C-6 Helldiver; Douglas A-26 Invader, JD-1 Invader, XC-112A, DC-6 (C-118,R6D) Liftmaster; Eastern XTBM-5 Avenger; Fairchild C-82 Packet, C-123B; Fleetwings BTK-1; Goodyear FG-1, 3, 4 Corsair; Grumman F6F Hellcat, F7F Tigercat, F8F Bearcat, AF-2W, S Guardian; Hughes D2A(XA-37); Lockheed RB-24 (PV-1) Ventura, PV-2 Harpoon, C-69E Constellation; Martin B-26 (JM-1,2) Marauder, AT-23 Marauder, PBM-5 Mariner, Model 202, A Mercury, 404 (RM-1); North American XB-28, AJ-1, 2 Savage; Northrop XP-56 Black Bullet, P-61 (F2T) Black Widow, F-15F Reporter; Republic P-47 Thunderbolt; Sikorsky S-56 (H-37,HR2S), HR2S-1W; Stroukoff Pantobase; Vickers Warwick I, II; VoughtAU-1, F4U Corsair, XTBU-1. | An 18-cylinder two-row radial engine providing up to 2,500 horsepower. An important factor of Allied air supremacy in World War II, the Double Wasp remained in production until 1960. |
| **R-4360 WASP MAJOR** | Aero Sud-Est SE-2010; Boeing XB-44 Super Fortress, B-50, C-97, Model 377 Stratocruiser, XF8B-1, B-377PG Pregnant Guppy; Consolidated Vultee B-36, XC-99; Convair A-41; Curtiss Wright XBTC-2; Douglas TB2D Devastator; C-74 Globe-master I, C-124 Globemaster II; Fairchild C-119 (R4Q) Packet, C-120 Packplane; Goodyear F2G Corsair; Hughes XF-11, HFB-1 Hercules; Lockheed R60-1 Constitution; Martin AM-1, 2 Mauler, JRM-2 Mars, P4M Mercator; Northrop B-35; Republic XP-72, XP-12 Rainbow. | A four-row, 28-cylinder radial with the cylinders in a spiral arrangement, providing 3,000 horsepower initially, and up to 4,300 horse-power in later models. |
| **J42** | Grumman F9F-2 Panther. | A centrifugal-flow turbojet developing 5,000 pounds of thrust. |
| **J-48** | Grumman F9F-5 Panther, F9F-6 8 Cougar; Lockheed F-94C Starfire; North American F-93A | A centrifugal-flow turbojet developing 7,250 pounds of thrust, up to 8,750 pounds with afterburner. |

| ENGINE MODEL | AIRCRAFT | NOTES |
|---|---|---|
| **CLASSIC ENGINES** | | |
| T-34 | Douglas C-133 Cargomaster; Boeing B-377SG Super Guppy, C-97J; Lockheed C-121F Constellation. | A turboprop developing 7,200 horsepower. |
| J-57 (JT3) | North American F-100 Super Sabre; McDonnell F-101 Voodoo; General Dynamics F-102 Delta Dagger; Boeing B-52 Stratofortress; Martin B-57D; Lockheed U-2, Boeing C-135A Stratolifter, KC-135A Stratotanker; Ling-Temco-Vought F8 Crusader; Douglas A3D Skywarrior, F4D, F5D Skyray; Boeing 707-120, 720, McDonnell Douglas DC-8-10; Northrop SM-62 Snark; Boeing C-137. | An axial-flow, dual compressor turbojet developing 13,500 pounds of thrust, up to 19,600 pounds with afterburner. |
| J75 (JT4) | Republic F-105 Thunderchief; General Dynamics F-106 Delta Dart; Martin P6M Seamaster; Lockheed U-2A; Boeing 707-220, -320; McDonnell Douglas DC-8-20, -30; North American F-107A. | An axial-flow, dual-compressor turbojet developing 17,500 pounds of thrust, up to 26,500 pounds with afterburner. |
| J52 | McDonnell Douglas A-4 Skyhawk series; Grumman A-6 Intruder series; North American Rockwell AGM-28 Hound Dog missile | An axial-flow turbojet developing up to 11,200 pounds of thrust. |
| J60 (JT12) | Rockwell International T-2B Buckeye, T-39A Sabre-liner; Lockheed C140 Jetstar. | A small, high-performance single-spool turbojet developing 3,300 pounds of thrust. |
| J58 | Lockheed SR-71 Blackbird, YF-12A. | A turbojet engine, in the 30,000-pound thrust class, powering Mach 3 aircraft. |
| JT3D (TF33) | Boeing 707-120B, 320B & C, 323C, 720B; McDonnell Douglas DC-8-50, -60; DC-8F Jet Trader; Boeing B-52H Stratofortress; Convair RB-57F Angel; Boeing C-135B Stratolifter, KC-135B Stratotanker; Lockheed C-141A Starlifter; Boeing VC-137B,C Presidential Plane. | A turbofan version of the J57 developing up to 21,000 pounds of thrust. |

| ENGINE MODEL | AIRCRAFT | NOTES |
|---|---|---|
| **CLASSIC ENGINES** | | |
| **TF-30** | General Dynamics F-111; LTVA-7A, B, C Corsair II; Grumman F-14ATomcat. | An axial-flow turbofan developing up to 25,100 pounds of thrust with afterburner. |
| **MODERN COMMERCIAL ENGINES** | | |
| **JT8D FAMILY** | Boeing 727, 737-100/-200; McDonnell Douglas DC-9; Boeing MD-80, Super 27 Reengining Program. | Introduced 1964 with the inaugural flight of Boeing's 727-100 aircraft. Today, the eight models that comprise the JT8D standard engine family cover the thrust range from 14,000 to 17,400 pounds and power 727, 737, and DC-9 aircraft. More than 11,800 JT8D standard engines have been produced, accumulating over one-half billion hours of service operation as of 2007.<br><br>A modern derivative, the -200 series, covers the 18,500 to 21,700 pound thrust range and is the exclusive power for the popular MD-80 series aircraft. Since starting service in 1980, more than 2,900 engines have been produced. |
| **JT9D FAMILY** | Boeing 747, 767; Airbus A300, A310; McDonnell Douglas DC-10. | The JT9D opened a new era in commercial aviation: the high-bypass-ratio engine to power wide-bodied aircraft. As Pratt & Whitney's first high-bypass-ratio turbofan, it introduced many advanced technologies in the areas of structures, aerodynamics, and materials to maximize fuel efficiency and component durability. |

| ENGINE MODEL | AIRCRAFT | NOTES |
|---|---|---|
| **MODERN COMMERCIAL ENGINES** | | |
| | | The JT9D family consists of three distinct series. The JT9D-7 engine covers the 46,300 to 50,000 pound thrust range, and the JT9D-7Q series has a 53,000 pound thrust rating. Later models, the -7R4 series, cover the 48,000 to 56,000 pound thrust range. For JT9D-7R4 twinjet installations, the engines are approved for 180-minute ETOPS (extended-range twin-engine operations). |
| **PW2000 FAMILY**<br> | Boeing 757; Ilyushin IL-96; C-17. | The PW2000 covers the thrust range from 37,000 to 43,000 pounds. Designed for the Boeing 757, PW2000s accumulated more than 26 million hours through March 2003. A military version of the PW2000, the F117-PW-100, is the exclusive powerplant for the U.S. Air Force C-17 Globemaster III military transport. The U.S. Air Force also selected the PW2040 to power the C-32A, the military version of the 757. |
| **PW4000 – 94-INCH FAN**<br> | Boeing 747-400, 767-200/-300, Boeing MD-11; Airbus A300-600, A310-300. | The PW4000 94-inch fan model is the first in a family of high-thrust aircraft engines. For twin-engine aircraft, the PW4000 is approved for 180-minute ETOPS (extended-range twin-engine operations).<br><br>Since entering service in 1987, the PW4000 has offered airlines excellent operating economics and high reliability. Advanced, service-proven technologies, such as single-crystal superalloy materials and its full-authority digital electronic control (FADEC), contribute to superior fuel economy and reliability. |

| ENGINE MODEL | AIRCRAFT | NOTES |
|---|---|---|
| **MODERN COMMERCIAL ENGINES** | | |
| **PW4000 100-INCH FAN** | Airbus A330-300, A330-200. | The PW4000 100-inch fan engine is the first derivative model in the PW4000 family. Developed specifically for the Airbus A330 twinjet, it is certified from 64,500 to 68,600 pounds of thrust. A PW4168A thrust rating is available with a 4.5% increase in thrust for high altitude and hot takeoff conditions. |
| **PW4000 112-INCH FAN** | Boeing 777-200/-300. | The PW4000 112-inch fan engine is the second derivative model in the PW4000 family. The PW4084, certified at 86,760 pounds thrust, was the launch engine for Boeing's 777 super twinjet. It entered service in June 1995 with United Airlines, already qualified for 180-minute ETOPS — an industry first. It is also the first engine to operate with approval for 207-minute ETOPS. |
| **V2500** | Airbus A319, A320, A321; Boeing MD-90. | International Aero Engines' V2500 turbofan engine family covers the 22,000 to 33,000 pound thrust range and provides efficient, clean power for more than 80 customers. The V2500's advanced technology makes it the engine of choice in its class, delivering unequaled efficiency and reliability. The engine's |

| ENGINE MODEL | AIRCRAFT | NOTES |
|---|---|---|
| MODERN COMMERCIAL ENGINES | | |
| | | wide-chord, shroudless fan blade design not only increases fuel efficiency, but also provides superior tolerance to foreign-object damage. The high-pressure compressor is a rugged 10-stage design with advanced airfoil aerodynamics, resulting in higher component performance. A two-stage high-pressure turbine provides excellent efficiency and long component life. Engine operability and maintenance diagnostics are enhanced by the full authority digital engine control. |
| PW6000 | Airbus A318. | The PW6000 is the most recent commercial product to be developed at Pratt & Whitney. It covers the 18,000 to 24,000 pound thrust class and has been designed specifically for 100-passenger aircraft. It is currently offered on the Airbus A318.<br><br>The PW6000 builds on proven technology gleaned from other Pratt & Whitney advanced engine programs to deliver the lowest cost of ownership for 100-passenger aircraft operators. Pratt & Whitney has incorporated technological advances in the PW6000 that enable a reduction in parts count. With fewer parts, the engine has a lower acquisition cost as well as a reduced maintenance cost. |

| ENGINE MODEL | AIRCRAFT | NOTES |
|---|---|---|
| **MODERN COMMERCIAL ENGINES** | | |
| **GP7000 FAMILY** | Airbus A380, A380F. | Two of the most respected engine manufacturers in the industry, Pratt & Whitney and General Electric, bring the new GP7000 engine family to A380 customers. Only the Alliance can offer a total business solution backed by world-class product support and service.<br><br>The GP7000 is derived from some of the most successful wide-body engine programs in aviation history—the GE90 and PW4000 families. These engines demonstrated industry-leading ETOPS reliability from service entry and forged a record of over 250 million hours of superior performance. Building on the GE90 core and the PW4000 low-spool heritage, the GP7000 is a refined derivative with an infusion of new technologies. |
| **MILITARY ENGINES** | | |
| **F100 FAMILY** | F-15, F-16. | Powering all of the U.S.Air Force's (USAF) F-15 fighter aircraft and the majority of F-16 fighters in 22 countries and the USAF, Pratt & Whitney's family of F100 engines is the mainstay of air forces world-wide. With more than 7,000 engines produced and over 21 million flight hours, the F100 fighter engine has a remarkable record of safety and reliability. |
| **F117** | Boeing C-17 Globemaster. | The F117-PW-100 engine is the military version of the PW2000 commercial engine. Certified at 40,400 pounds of thrust, the F117 was selected by the U.S. Air Force as the exclusive powerplant for the C-17 Globemaster III, an advanced four-engine transport. |

| ENGINE MODEL | AIRCRAFT | NOTES |
|---|---|---|
| **MILITARY ENGINES** | | |
| | | Unique to the C-17, the F117 engines are equipped with a directed-flow thrust reverser capable of being deployed in flight. On the ground, the thrust reverser can back a fully loaded aircraft up a two-degree slope. It is also noteworthy that the F117-powered C-17 set 22 world records during qualification testing before initial operating capability. |
| F119 | F-22 Raptor. | Pratt & Whitney's F119 turbofan engine, the world's most advanced aircraft engine in production, meets the need for greater speed and lower weight for new military weapon systems. In the 35,000 pound thrust class, the engine is a dual-spool, counter-rotating turbofan that enables aircraft operation at supersonic speeds for extended periods without thrust augmentation. |
| F135 | F-35 Joint Strike Fighter. | Pratt & Whitney's F135 propulsion system is the power of choice for an advanced, single-engine tactical fighter, the F-35 Joint Strike Fighter (JSF), being developed by Lockheed Martin. The F-35 has unique capabilities for land-based conventional takeoff and landing (CTOL), carrier-variant takeoff and landing (CV) , and short takeoff and vertical landing (STOVL). <br><br> The F135 propulsion system has proved that it can meet these diverse requirements. The F135 is an evolution of the F119-PW-100, a technologically advanced turbofan that powers the Air Force's F-22 Raptor. |

151

| ENGINE MODEL | AIRCRAFT | NOTES |
|---|---|---|
| **PRATT & WHITNEY CANADA** <br> **TURBOPROP** | | |
| **PT6A TURBOPROP FAMILY** <br><br> **(MORE THAN 50 MODELS OF THE PT6 ARE REPRESENTED IN THE NEXT COLUMN)** | Air Tractor AT 402A/402B, 502B, AT 602, T802/802A/802AF/802F; AMI DC-3; Ayres Model 660, Turbo Thrush T-65, T-34, T-15; CATIC/HAIG Y-12; Basler Turbo BT-67; Blackhawk XP™ King Air 90 series, Blackhawk 200xp™, 425xp™, Cheyenne XP™; Blue 35; Cessna 208/208B Caravan I, Cessna Conquest I; Conair Aviation – S2 Turbo-Firecat; deHavilland DHC-6 Twin Otter Series 300, Dash 7, DHC-4 EADS Socata TBM 850, 700; Embraer Bandeirante EMB-110, EMB-111, EMB-121 Xingu, EMB-312 Tucano, EMB-314 Super Tucano, Embraer Caraja; Frakes Mallard, Turbo Cat Model A/B/C; Ibis Aerospace Ae 270 HP; JetPROP DLX; KA1-KT-1/KO-1; LET L410; National Aerospace Laboratories Saras; Pacific Aero Cresco 750, 750XL; PIAGGIO P-166-DL3, Avanti P-180,Avanti II; Pilatus Turbo Porter PC-6, Turbo Trainer PC-7/PC-7 MKII, Turbo Trainer PC-9, PC-12, PC-21; Piper Cheyenne IA, II/IIXL, III/IIIA, Piper Malibu Meridian, Piper T1040; Polish Aviation Factory M28 Skytruck; PZL-Okecie PZL-130 TE Turbo-Orlik, TC-II Turbo-Orlik, PZL-106 Turbo-Kruk ; Quest Kodiak; Raytheon Beech 1900/1900C, 1900D, 99, 99A, B99, C12F, C99 Air-liner, King Air A100, King Air C90A/B/SE,E90, F90-1/C90GT, RC-12K, Starship, King Air 200/B200, King Air 300/350, T-34C, T-44A, T-6A Texan II; Reims F406 Caravan II; Schweizer G-164B AG-Cat Turbine; G-164D AG-Cat Turbine; Shorts 330, 360/360-300, C-23A Sherpa, C-23B Super Sherpa; Silverhawk 135; Socata TBM 700 T-G Aviation Super Cheyenne®; Turbine Air Bonanza; Vazar Dash 3 Turbine Otter. | The PT6A engine family has been the backbone of small turboprop aircraft for decades. With a low cost of operation, backed by an extensive field support network, the PT6A has also been providing power to a wide range of turboprop aircraft, both twin and single. With its legendary performance track record, the PT6A engine has emerged as the undisputed leader in the general aviation market. <br><br> For nearly four decades, the PT6A has set the industry benchmark for reliability and dependability. Ranging from 580 to 1,940 shaft horsepower, it offers the best value available, along with simplicity of design, ease of maintenance, and efficiency of operation. |
| **PW100 FAMILY** | Alenia Aeronautica EADS ATR 42-300/320; 42-400/500; 72-200 72-210/500;Alenia Aeronautica EADS – CASA C-295 XIAN Aircraft Co. MA-60 Canadair CL-215T/CL-415 Bombardier Aerospace Q100, Q200, Q300, Q400; Fairchild Dornier 328-110/120; Embraer EMB120 Brasilia; Fokker 50/High Performance; Fokker 60 Utility; Ilyushin IL-114-100; Jetstream Aircraft ATP. | The PW100 is an advanced technology, fuel-efficient turboprop engine designed to power 30- to 70-passenger regional transport, as well as utility and corporate aircraft. <br><br> The PW100 engine family has grown to cover a wide range of power from 1800 shaft horsepower to 5000 shaft horsepower. |

| ENGINE MODEL | AIRCRAFT | NOTES |
|---|---|---|
| | **TURBOSHAFTS** | |
| PT6 AND PW200 FAMILIES (HELICOPTER ENGINES) | Agusta A119 Koala PT6B-37A; Agusta Bell PT6T-3/6; Agusta Bell 412/SP/HP/EP PT6T-3B/BE/D/DF/6/B; AgustaWestland AW139 PT6C-67C; Bell 212 PT6T-3/B; Bell 412/SP/HP/EP PT6T3B/BE/BF/BG/D/DF; Bell CFUTTH CH-146 Griffon PT6T-3D; Bell UH-1N/AH-1J/CUH-1N T400-CP-400; Bell AH-1T/AH-1J T400-WV-402; Dyncorp UH-1 Global Eagle PT6C-67D; Bell VH-1N T400-CP-401; Sikorsky S-58T PT6T-3/6; Sikorsky S-76B PT6B-36A/B; Agusta A109E Power PW206C; Agusta A109 Grand PW207C; Bell 427 PW207D; Bell 429 PW207D1; Eurocopter EC135 PW206B/B2; Kazan Ansat PW207K; MD Explorer. | P&WC has a 30-year history in providing helicopter power, and the PT6 turboshaft family continues to set standards for helicopter engine reliability, durability, and low ownership cost. The PW200 engine delivers the latest proven technology leading to unprecedented levels of reliability and economy for the new generation of single- and twin-engine helicopters and is an excellent engine for rotorcraft and tilt-rotor UAVs. |
| | **TURBOFANS** | |
| JT15D, PW300, PW500 AND PW600 FAMILIES | JT15D family<br>Aerospatiale Corvette; Agusta S211/S211A; Cessna Citation I, II/S II, V, Ultra, Cessna UC-35A/B; Mitsubishi Diamond IA; Raytheon Hawker 400XP; Raytheon T-1A Jayhawk; Raytheon TCX. | The JT15D develops thrust in the 2,200 pounds to 3,350 pounds range. JT15D engines offer the added benefit of engine commonality across a range of business jet models for ease of maintenance. |

| ENGINE MODEL | AIRCRAFT | NOTES |
|---|---|---|
| | **TURBOFANS** | |
| PW300 family<br>Bombardier Learjet Model 60; Cessna Citation Sovereign; Dassault Falcon 2000EX , 7X Fairchild Dornier 328JET; Gulfstream G200; Learjet Model 60; Raytheon Hawker 1000, Hawker 4000. | | The PW300 family incorporates the latest innovations in component technology to provide unparalleled benefits for super mid-size corporate aircraft.<br><br>The PW300 offers a series of advanced, high bypass ratio turbofan engines, in the 4,500 pounds to 8,000 pounds thrust range, designed for clean, quiet, and low-cost operation.<br><br>Designed using the latest component technology, the PW300 family offers competitive fuel efficiency and optimum thrust-to-weight ratios. |
| PW500 family<br>Cessna Citation Bravo, Excel, XLS, Ultra Encore, Encore+ , Cessna UC-35 C/D PW535A. | | The PW500 turbofan family incorporates many of the latest advances to offer a highly reliable, quiet, and cost-effective power solution.<br><br>The PW500 is an advanced, high by-pass ratio turbofan engine in the 3,000 pounds to 4,500 pounds thrust range that offers a common core concept for a wide range of aircraft requirements. It incorporates improved component technologies to provide new standards of durability and reliability. |

| ENGINE MODEL | AIRCRAFT | NOTES |
|---|---|---|
| **TURBOFANS** | | |
| | PW600 family<br>Cessna Mustang PW615F-A.<br>Eclipse 500 PW610F-A. | Environmental friendliness, low fuel consumption, high durability and reliability, and low maintenance cost are all hallmarks of the new PW600 family.<br><br>The PW600 was specifically designed for the business and general aviation markets.<br><br>The PW600 has been designed to deliver a wide range of thrust while delivering the highest standards in reliability and durability. The ability to scale the engine to meet a wide range of customer requirements has been a key driver in the PW600 design, and sets the stage for the new family of turbofan engines in the 900 to 3,000 pounds thrust range. |
| **AUXILLARY POWER UNIT** | | |
| PW900<br> | APU power for the Airbus A380 and Boeing 747-400. | The PW900 is an auxiliary power unit (APU) designed and developed to provide pneumatic power for main engine start and cabin environmental control. It also has drive pads for two airframe-supplied generators that provide electrical power to the airplane. |
| **ROCKET POWER (DOES NOT INCLUDE ROCKETDYNE ACTIVITIES PRIOR TO ACQUISITION BY PRATT & WHITNEY)** | | |
| SSME<br> | Space shuttle. | The space shuttle main engine (SSME) is the world's most reliable and highly tested large rocket engine. The SSME has achieved 100% flight success with a demonstrated reliability exceeding 0.9996 in over 1,000,000 seconds of hot-fire experience. |

| ENGINE MODEL | AIRCRAFT | NOTES |
|---|---|---|
| **ROCKET POWER (DOES NOT INCLUDE ROCKETDYNE ACTIVITIES PRIOR TO ACQUISITION BY PRATT & WHITNEY)** | | |
| **RS-68**<br> | Delta IV family of evolved expendable launch vehicles (EELV). | The RS-68 engine is the first new large liquid-fueled rocket engine to be developed in the United States in 25 years.<br><br>The bell-nozzle RS-68 is a liquid hydrogen–liquid oxygen booster engine utilizing a simplified design philosophy resulting in a drastic reduction in parts compared to current cryogenic engines. This design approach results in lower development and production costs. |
| **RS-27A**<br><br>Image courtesy of United Launch Alliance | Delta II launch vehicle. | The nominal 200,000 pound sea-level thrust RS-27 has compiled one of the most consistent and successful launch records in the history of rocketry with a 100% reliability factor.<br><br>A single-start powerplant, it is gimbal-mounted and operates on a combination of liquid oxygen and RP-1 (kerosene). The thrust chamber is regeneratively cooled with fuel circulating through 292 tubes that comprise the inner wall of the chamber. |

| ENGINE MODEL | AIRCRAFT | NOTES |
|---|---|---|
| **ROCKET POWER (DOES NOT INCLUDE ROCKETDYNE ACTIVITIES PRIOR TO ACQUISITION BY PRATT & WHITNEY)** | | |
| **RD-180**<br><br><br><br>Image courtesy of Lockheed Martin Corporation | Lockheed Martin Atlas III and V. | Packing nearly one 1,000,000 pounds of thrust in a total propulsion system, the RD-180 delivers a 10% performance increase over current operational U.S. booster engines. |
| **RL-10**<br><br><br><br>Image courtesy of United Launch Alliance | Atlas and Delta launch vehicles. | As the most reliable, safe, and high-performing upperstage engine in the world, the remarkable RL-10 has accumulated one of the most impressive lists of accomplishments in the history of space propulsion.<br><br>Created in 1959 after Pratt & Whitney harnessed high-energy liquid hydrogen as fuel for aerospace propulsion, the RL10 has helped place numerous military, government, and commercial satellites into orbit and powered space probe missions to nearly every planet in our solar system. |

# Index

1830-C Twin Wasp, 23–24
40-A, 16

## A

A330, 118–119
A340, 118
Achieving Competitive Excellence (ACE), 126–128
Active clearance control, 104
Adelman, Barney, 48
Advanced Tactical Fighter (ATF), 98, 111
Advanced Turbo Prop (ATP), 88
AE2100 turboprop, 88
AE3007 turbofan, 88
Aerospace, 47–48, 52–53, 56–59
Airbus A310, 107–108
Airbus A318, 129
Airbus Industries, 75–76
Air Commerce Act, 16
Air-cooled engines, 6–7
Aircraft carriers, 4
Airfoils, 104
Air Force, 92
    F-15/F-16, 93–97
    FX program, 93
Airlift, 71
Airline deregulation, 87, 101
Airmail, 16–17
Air Mail Act, 16
Air Mail Scandal, 19–20
Air racing, 17–18
Air taxi market, 90
Allison, 88
Allison T56, 42
Altman, Dave, 48
Andrew Willgoos Gas Turbine Laboratory, 36
Apollo, 58, 59

ARAMCO pipeline, 60
Arms Deal of the Century, 97
Arnold, Hap, 22
Arnold Engineering Development
    Center (AEDC), 115
Atomic Energy Commission, 56
Atomic power, 56
Automotive industry, 2, 27
Auto racing, 84–86
Auxiliary power units, 84
Aviation industry
    beginnings of, 16
    pre-WWII, 25
Axial-flow engines, 36, 40

## B

B-2, 42
B-24 Liberation, *32*
B-35 Flying Wing, 39
B-47, 42
B-50, 39
B-52 bomber, *38*, 42, 44
Beech Expeditor, 83
Berlin Airlift, 39
Black, Hugo, 19
Boeing, William, 11, 20
Boeing 247, 18–19, *19*
Boeing 707, 44
Boeing 717, 129
Boeing 727, 69–70, *71*
Boeing 737, 101–102
Boeing 747, 73–75
Boeing 757, 101, 102, 104–105
Boeing 777, 119–122
Boeing Aircraft of Canada, 11
Boeing Airplane Company, 11
Boeing Air Transport, 11, 16

159

Note: Locators in *italics* indicate figures.

Boeing B-50, 39
Boeing F2B, 9
Borrup, Jack, 3, 6, 8, 14
Boyne, Walter, 26
Bristol Engine Company, 46
Brooks, Carroll, 12, 14
Brown, Bill, 54
Brown, Don, 3, 6, 8, 21, 22
Brown, Walter, 19
Bruner, Jim, 107
Buick, 27
Burton, David, 30

**C**

C-5A, 71, 73
C-17, 105
C-54, 39
C-124 Globemasters, 39
C-130, 71
C-141, 71
Canadian market. *See* Pratt & Whitney Canada
Canadian Propellers Ltd., 78
Carlson, Robert, 123
Cessna, 86, 88–89
CFM56, 102
CFMI, 100, 136
Chance Vought Corporation, 11
Chemical Systems Division, 48–50
Chênevert, Louis, 130
Chevy, 27
CJ805, 45
Clarkson, Larry, 99–100, 107
Clipper services, 17
Coar, Dick, 96
COIN (counterinsurgency operations), 84
Cold War, 39, 56, 124
Comet, 37, 45–46
Commercial air transport, emergence of, 15–16
Commercial aviation, 69–76, 118
Commuter/regional market, 87–88
Computer-aided design, 104
Concorde, 73
Connecticut Automated Nuclear Engine
  Laboratory (CANEL), 56
Continental Motors, 27
Convair 880, 45
Convair B-36, 39
Convair Centaur, 52
Conventional takeoff and landing
  (CTOL), 116
Conway engines, 45–46
Coran, Mark, 124

Corsair, 39
Coyote Canyon, 49
Crane, Henry, 3
Crane-Simplex company, 3
Cullen, James, 5
Curtiss Hawk, 9

**D**

David, George, 124
Deeds, Edward, 8
Defense budgets, 15
deHavilland, 45, 84
"Dependable Engines" motto, 10
Depression, 15, 19
DH-4, 16
Direct-cycle engine, 56
Dogfighting, 92–94
Double Wasp, 22, 26, 30, 32, 39
Douglas, Donald, 19
Douglas C-74, 39
Douglas DC-2, 19
Douglas DC-3, 19
Douglas DC-4, 39
Douglas DC-8, 44
Douglas DC-9, 70
Douglas DC-10, 74–75
Drummond, John, 78, 79–80

**E**

Earhart, Amelia, 17
East Hartford plant, 10–11, 21, 26, 50
EcoPower wash, 136
Electricity generation, 56–59
Engine Alliance, 100, 128–130
Engine companies
  *See also specific companies*
  partnerships among, 99–100, 128–130
Engine fans, 120
"ETOPS Out of the Box" certification, 118–119, 122
Extended twin-engine operations (ETOPS), 118–119

**F**

F-100, 93–94, 97, 100
F-111, 92
F-111B, 92
F117, 105
F119, 98, *112*, 114–115, 116, 117
F135, 98
F-4, 93
F-15, 93–97
F-16, 93–97
F-22, 98, 114–115

F-35, 98, 117–118
F401, 93–94
Fan blades, 120
Farmer, Tom, 113, 115, 117
Ferry Command, 79
Florida Research and Development Center, 51–55, 125
Ford Motors, 27
Forrestal, James, 31
France, 25–26
Frosech, Charles, 69
FT4, 60
FT8, 60–61
Fuel burn, 129
Fuel Cell Operations, 58–59
Fuel cells, 56–59
Full-authority digital electronic engine controls (FADEC), 104

**G**

Gas turbine engines, 34, 36–37, 59–61
Gee-Bee racer, *23*
GE37, 111
Gearboxes, 131–132
Geared Turbofan engine (GTF), *126*, 130–133, *134*
General Electric (GE), 40, 44–45
    Airbus and, 76
    alliance with, 128–130
    competition with, 111–115
    direct-cycle engine and, 56
    Great Engine War and, 96–98
    high-bypass designs of, 73
    industrial segment of, 60
    J79, 91
    SNECMA and, 100, 102
    turbofans, 88
Germans, 26
Gillette, Frank, 111–112
Global Service Partners, 136
GP7000, 128–129
Granatelli, Andy, 84–85
Gray, Harry, 20, 106
Great Britain, 26, 37
Great Engine War, 96–98
Grumman Hellcat, 29
Gunston, Bill, 44
Gwinn, William, 22, 33, 52

**H**

Hamilton Standard (HSD), 21, 87, 104
Hamilton Standard Propeller Corporation, 11
Hartford, Connecticut, 4–5, 7–8
Heron, Sam, 3

High-bypass designs, 73
Hobbs, Leonard "Luke," 21
Horner, H. Mansfield "Jack," 12, 18, 22
    Hughes and, 40
    on jets vs piston engines, 34–36
    on the JT3-A, 44
Hornet, 9, 10, *16*
Hughes, Howard, 17, *18*, 40
Hydrogen fuel, 50–51, 52
Hydromechanical controls, 104

**I**

IL96M program, 105
Indirect cycle, 56
Industrial gas turbines, 59–61
International Aero Engines (IAE), 100, 109
International Civil Aviation Organization (ICAO), 119
International partners, 97–98
Ito, Yuzuri, 125
Ito University, 125

**J**

J42, *37*
J47, 40, 44
J52, 69, 69–70, *91*
J57, 34, *38*, 44, 91
J58, 50, 53–54
J75/JT4, 44, 45, 91
J79, 44–45, 91
J91, 56
Jacobs Aircraft, 27
Japanese, 26, 109
Jet fuels, 102
Jets, 33–34, 40–46
    commercial engines, 69–76
    vs piston engines, 34–36
    regional, 88
    VLJs, 89–90
Johnson, Kelly, 50
Joint Advanced Fighter Engine, 111–115
Joint Strike Fighter (JSF), 115–118
JSF119, 116–118
JT10D, *99*, 100–102
JT15D, 86
JT3, *41*, 44, 45
JT3-A, 42, 44
JT3D, 45, 73
JT3/JT4 family, 59
JT8D, 60, 69, *70*
JT8D-200, *70*, 71, 100

JT9D, *73*, 73–75, 106
JTF10, 91
JTF10A, 69
JTF17, 55

**K**
Kansas City plant, 27, 29
Kevlar, 120
King Air, 84, 86
Knudsen, William, 27, 29
Korean War, 79
Krapek, Karl, 124, 125
Kuter, Lawrence, 73

**L**
Lampert Committee, 15
Lawrance, Charles, 3
Lawrance J engine, 3
Lawrence, Herb, 48
Layoffs, 125
Lean manufacturing, 126–128
Leighton, Bruce, 7
Liberty engine, 26
Liftfan design, 116–117
Lightning II, 115–118
Lindbergh, Charles, 3, 12, 16
Line replaceable units (LRUs), 113
Liquid-cooled engines, 6–7, 20, 22
Liquid-fueled engines, 52–53
Liquid-hydrogen engines, 133
Little Pratt. *See* Pratt & Whitney Canada
Lockheed F104, 44
Lockheed L-1011 Tristar, 73, 84
Lucky Lady II, 39

**M**
Mach 3, 54
Machine tool industry, 4–5
Maintenance business, 135–136
Maintenance, repair, and overhaul
  business (MRO), 135
Market feedback analysis, 127
Marks, Charlie, 3, 6, 8
McNamara, Robert, 92
Mead, George, 3, 6, 8, 21, 27
Middletown, Connecticut plant, 55–56, *72*
MiG-23 Flogger, 93
MiG-25 Foxbat, 93
Military spending, decline in, 123–124
Mills, Eunice, 29
Missileer, 91–92

Missimer, Bill, 96–97
Module centers, 126
Moffett, William, 4
Morrow Board, 15
Mulready, Dick, 50–51

**N**
NASA, 58–59
Nash-Kelvinator, 27
NATO fighter program, 97, 100
Navy, 91–92, 93
Nene engine, 36–37
New Deal, 15, 19
Newland, Allan, 82
Niles -Bement-Pond company, 4, 6
Northrop Aircraft, 11
Nuclear engines, 56

**O**
O'Connor, Jim, 114
Oil pipelines, 60
Olympus engine, 46
"One-piece flow" manufacturing, 124–125

**P**
Pacific Air Transport, 11
Pan American, 17, 73, 107–108
Parkins, Wright, 24, 47, 51, 82
Piston engines, 34–36, 37–40
Pizzi, Tony, 112
Podolny, Bill, 59
Post, Wiley, 17
Pratt & Whitney, 11
  in 1930s, 15–24
  alliances of, 100–102, 108–110, 128–130
  beginnings of, 7–9
  challenges for, at end of 20th century, 123–128
  commercial aviation and, 69–76
  competition with GE, 111–115
  founding of, 4–7
  fuel cells and, 56–59
  Great Engine War and, 96–98
  growth of, 10–12
  industrial gas turbines of, 59–61
  innovations by, 30–31
  maintenance business of, 135–136
  move to jet engines by, 40–46
  during post-war years, 33–46
  restructering of, 124–128
  space race and, 47–53
  timeline, 62–67

during WWII, 25–32
Pratt & Whitney Canada, 11, 77–90
    commuter/regional market and, 87–88
    engines of, 88–90
    manufacturing by, 78–80
    PT6, 81–84, 84–86
    service business of, 77–78
    in surplus business, 79
    TDEs of, 87
Pratt & Whitney Machine Tool, 5, 6
Pratt & Whitney Power Systems, 60–61
Price competition, 99
"Proof-of-concept" engine, 87
PT-1, 34
PT4/T57, 42
PT6, 81–84, 84–86, 87
Putt, Donald, 48
PW100, 87–88
PW150, 87, 88
PW200, 88
PW300, 88–89
PW500, 88–89
PW600, *89*, 90
PW800, 88
PW900, 89
PW2000, *103*
PW2037, 102–105
PW4000, *106*, 106–108, *117*, 119
PW4084, 119–122
PW4168, 119
PW5000, 111–115
PW6000, 88, 129–130

## Q

Quality initiatives, 126–128

## R

R-1830 Twin Wasp, 22
R-2000 Twin Wasp, 22
R-2800 Double Wasp, 22, 39
R-3350 Turbo-Compound, 39
Racing engines, 84–86
RB432, 109
RD-180, 133
RD-AMROSS, 133
Regional jets, 88
Regional market, 87–88, 129
Renegotiation Act, 31
Rentschler, Frederick Brant, 43
    on Air Mail Scandal, 20
    background of, 2–4

death of, 46
founding of Pratt & Whitney by, 4–7
on piston engines, 38
on postwar readjustment, 31
vision of, 1–2
at Wright Aeronautical, 2–3
Rentschler, George, 2, 5
Rentschler, Gordon, 2, 4
Riley, Ron, 79
RL10, 52, *54*, 133
RL20, 52
Robertson, Bill, 107
Rocketdyne, 133–135
Rocket engines, 48–50, 52–53, 133–135
Roelke, Chuck, 51
Rolls-Royce
    Airbus and, 76
    alliances with, 100–102, 108–110
    Conway engines, 45–46
    high-bypass designs of, 73
    Nene engine, 36–37
Roosevelt, Franklin D., 19, 25
Rosati, Bob, 31, 105, 108, 109
Rowe, Brian, 96
Russian aircraft industry, 105, 133
Ryder, Erle, 7

## S

Shamrock, 50
Shingijutsu, 124–125
Short takeoff and vertical landing (STOVL), 116–117
Sikorsky Aviation Corporation, 11
Silva, Rick, 112, 113
Single-crystal turbine blades, 104
Slay, Alton, 94
Smith, Elvie, 83
SNECMA, 76, 100, 102
Solid rocket motors, 48–50
Sorsensen, Charles, 27
Soviet aircraft, 93
Space race, 47–53
Space shuttle, *130*
Space shuttle main engine (SSME), 52–53
Space Shuttle Orbiter, 59
Spoils Conference, 19
SR-71 Blackbird, 53, 54
Stage III noise rules, 71
Standard work, 130
Steaman Aircraft, 11
Stephenson, Thor, 83n2
Stout Air Services, 11

STP, 84–85
Stratocruiser, 39
Stratofortress, 42
Subcontractors, 27
Suntan, 50, 52
Supercooling, 116
Supersonic transport (SST), 55
Super Sabre, 44
Surplus market, 39–40, 79

**T**
T34, 40
Tactical Fighter Experimental (TFX), 92
TALON, 119
Taylor, Philip, 36
TDE-1, 87
Technology demonstration engines (TDEs), 87
Technology readiness, 130
TF30, 91, *93*
TF33, 44
TF39, 73
Timeline, 62–67
Titan III family, 49
Torrell, Bruce, 83, 101
Trains, 84
Trippe, Juan, 17, 73
Turbine cars, 84–86
Turbine engines, 39, 40
Turbofans, 45, 71, 73, 86, 88, 88–89
Turbo Power & Marine (TP&M), 58, 60–61
Turboprops, 42, 81–84, 87–88
Turbo Train, 84
Turner, Roscoe, *17*
TWA, 40
Twin engines, 118–120
Twin Otter, 84
Twin-spool design, 42
Twin Wasp, 10, 22, 23–24, 32

**U**
U-2, 53
United Aircraft and Transport (UA&T),
   11–12, 16–17, 19
United Aircraft Corporation (UAC), 20, 21
United Airlines, 17
United Research Corporation, 48
United Space Boosters (USBI), 49

United States Air Mail Service, 16
United States Auto Club (USAC), 85
United Technologies, 20
United Technology Company, 48
*USS* Lexington, 4, 9
*USS* Saratoga, 4, 9
UTC Power, 59

**V**
V2500, 101, 109–110, 129
Variable compressor stators, 45
Very light jets (VLJs), 89–90
Vietnam War, 92–94
Vought, Chance, 11
Vought O2U-I Corsair, 9

**W**
War Production Board, 27
Washington Naval Limitations Treaty, 4
Wasp, *5, 7, 15, 16*
   in Boeing 247, 18
   in Canada, 79–81
   C models, 29
   designing of, 8–9
   R-1340, 10
   success of, 9–10
   during WWII, 26
Wasp Junior, 21
Wasp Major, 22, 32, *33*, 39
Water injection, 30
Wegner, Art, 114
Westinghouse J30, 34
Whirlwind, 3
White, Graham, 2
Whittle, Frank, 33
Willgoos, Andy, 3, 6, 8
Wilson, Eugene, 4
Women, in war years, 28, 29
Workforce, 29
World War I, 2
World War II, 25–32, 78–79
Wright Aeronautical, 2–3
Wright-Martin plant, 2

**Y**
Yellow Jacket, 22n3
Young, James, 77, 78, 79